Permanent-Magnet Materials and their Applications

Permanent-Magnet Materials and their Applications

K.H.J. Buschow

Van der Waals-Zeeman Institute, University of Amsterdam
Valckenierstraat 65, 1018 XE Amsterdam, The Netherlands

ttp TRANS TECH PUBLICATIONS LTD
Switzerland • Germany • UK • USA

Volume 5 of
Materials Science Foundations
ISSN 1422-3597

Distributed *in the Americas by*

Trans Tech Publications Inc
PO Box 699, May Street
Enfield, New Hampshire 03748
USA

Phone: (603) 632-7377
Fax: (603) 632-5611
e-mail: ttp@ttp.net
Web: http://www.ttp.net

and worldwide by

Trans Tech Publications Ltd
Brandrain 6
CH-8707 Uetikon-Zuerich
Switzerland

Fax: +41 (1) 922 10 33
e-mail: ttp@ttp.net
Web: http://www.ttp.net

Printed in the United Kingdom
by Hobbs the Printers Ltd,
Totton, Hampshire SO40 3WX

PREFACE

Permanent magnets play a role of increasing importance in modern society because they serve as vital components in numerous domestic and industrial devices. The last few decades have witnessed quite an extraordinary development in hard magnetic materials, especially after the advent of rare earth permanent magnets. The rare earth permanent magnets are unequalled because they combine a high magnetization with an extraordinary magnetic hardness which allows utilisation of such magnets in extreme shapes and small dimensions as required in modern devices. Magnets based on Sm and Co are unsurpassed with regard to their low temperature coefficients of coercivity and magnetization, even at temperatures far above room temperature. Magnets based on Nd, Fe and B have led to record maximum energy products and owe their more wide-spread utilisation to their comparatively low price. In the present report a review will be given of the progress made in the past years, including novel developments in the manufacturing routes and the physical principles on which these new developments are based. This includes a discussion of permanent magnets with remanence enhancement and permanent magnet materials based on interstitially modified alloys. The traditional permanent magnets based on ferrites, Al-Ni-Co alloys and Fe-Cr-Co alloys are still of commercial importance because they are considerably less expensive, although some of their magnetic properties are much inferior than those of rare earth permanent magnets. Other inexpensive permanent magnet alloys that will be discussed in this report, such as Mn-Al alloys, are still utilized but only in special application and on a very modest scale. Permanent magnets based on noble metals such as Pt-Co and Pt-Fe alloys combine fairly good hard magnetic properties with high mechanical strength and high corrosion resistance, but their high costs imply limited utilization as for instance in biomedical applications.

Professor K.H. Jürgen Buschow teaches solid state physics at the Universities of Amsterdam and Leiden. His main interest is in physical properties of intermetallic compounds. Current activities include research on various types of magnetic materials, such as magnetically hard and soft materials and heavy-electron systems. Important aspects of the research activities comprise investigations of the nature and strengths of magnetic interactions of systems composed of localised rare earth moments and itinerant 3d moments, 3d anisotropy and crystal field-induced 4f moment anisotropy. He has many years experience in metal hydride research and permanent magnet materials research performed during his previous employment at the Philips Research Laboratories in Eindhoven where he held the position of Chief Scientist. He has published more than 1000 research papers and several book chapters and review articles. He is Editor-in-Chief of the Journal of Alloys and Compounds and Advisory Editor of the Journal of Magnetism and Magnetic Materials.

1. INTRODUCTION

The first permanent magnet material that became available in sufficiently large quantities was carbon steel, developed during the last part of the 19th century. It was soon followed by tungsten steel with a slightly better performance. A considerable advancement was reached with the introduction of Co-steel or Honda steel in which about 35% of Co was substituted for Fe in Fe-W-C steel. However, because of the high price of Co compared to Fe, technological application of this new type of permanent magnet material proceeded relatively sluggishly.

The next step forward was achieved in 1932 by the introduction of MK steel by Mishima. This material was based on an alloy of Fe and Ni with small amounts of Co and Al. Its main advantages were that its price was only about one third that of Honda steel and that its performance as a permanent magnet material was better than Honda steel. This beneficial result was primarily a metallurgical achievement reached by means of a precipitation hardening process. Later investigations revealed that the alloy compositions correspond to regions of complete solid solubility. However, the equilibrium conditions at low temperatures correspond to a two-phase region and contain an Fe-rich bcc phase as well as a NiAl-rich bcc phase. Cooling of the homogeneous alloys to lower temperatures therefore involves phase separation into the two bcc phases. The improvement in hard magnetic properties involves mainly the formation of a specific microstructure reached after an equally specific heat treatment.

In fact, the magnets of MK steel can be regarded as precursors of the well-known alnico magnets developed only a few years later, in 1936. The compositional freedom and the possibility to optimise the heat treatment of this alloy system eventually led to the development of the well-known Ticonal G. The microstructure of these latter magnets was no longer isotropic but anisotropic. One of the essential processing steps consisted of a heat treatment in the presence of magnetic field by means of which columnar growth of the Fe-rich particles along a <100> direction could be achieved. A further improvement was reached in 1949 by applying a specific process of solidification of the alnico melt making it possible reach a certain degree of grain orientation. These grain oriented ingots were subsequently subjected to a similar heat treatment as mentioned above, which finally resulted in the magnet material formerly referred to as Ticonal GG and eventually to Ticonal XX in 1956. In the Alnico-type magnets mentioned above the magnetic anisotropy originates from the typical shape of the Fe-rich precipitates present in the microstructure.

The maximum magnetic anisotropy in all these materials is determined by the difference in demagnetizing factor between the long direction of the Fe-rich precipitates and their short direction. Much larger magnetic anisotropies can be obtained in materials having a strongly anisotropic crystal structure, such as found in crystal structures of hexagonal or tetragonal symmetry. Such materials were discovered already in 1936, when it was found that in tetragonal CoPt the coercivity was fairly high owing to the presence of a high intrinsic magnetocrystalline anisotropy. However, price considerations prevented widespread application of CoPt type magnets.

Ferroxdure is further example of a permanent magnet material exhibiting a high magnetocrystalline anisotropy. It is a ceramic material that can be described by the general formula MO $(Fe_2O_3)_6$, where M is one or more of the divalent metals Ba, Sr or Pb. The preparation of ferrite magnets is relatively easy and consists of comminution to particles of about 1 mm in diameter followed by sintering. The ferrites are ferrimagnetic and consist of several Fe sublattices with mutual antiparallel moment alignment. This is the reason that their magnetization is comparatively low. Nevertheless, the ferrites have found, and still find, widespread application owing to the low cost.

The hexagonal rare earth cobalt compounds of the type RCo_5 made their entry as permanent magnet materials in the 1960s when Hubbard et al.[1] found that the compound $GdCo_5$ displayed quite extraordinary hard magnetic properties. Initially, the interest in this compound remained fairly modest in view of the low magnetisation of this compound, which is associated with the antiparallel coupling between the Gd and Co sublattice moments. No significant magnetic hardness was found in several of the light rare earth compounds that had a considerable higher magnetization, for instance $NdCo_5$. This is probably the reason why it took several years before Strnat et al. [2,3] in 1966 discovered the excellent magnetic properties of YCo_5 comprising a huge magnetocrystalline anisotropy, high Curie temperature and high magnetization. Initially it was thought that the extremely large magnetocrystalline anisotropy was due to the cobalt sublattice. For this reason efforts to prepare permanent magnet bodies with large coercivities and large energy products were focussed on compounds like YCo_5 and $LaCo_5$ where the R component is non-magnetic. By means of a trial and error procedure Velge and Buschow [4,5] finally found that of all the RCo_5 compounds that of samarium is the most suitable for permanent magnet applications.

It has to be borne in mind that at that time crystal field theory was already well known among low-temperature physicists, but had not yet penetrated the community of material scientists working on permanent magnet materials. Therefore it was understood only several years later why a considerable advantage could also be derived from the anisotropy of the R sublattice.

The R_2Co_{17} compounds are richer in cobalt and therefore have a higher saturation magnetization. However, the cobalt sublattice anisotropy in such compounds proved to be not uniaxial. A compromise between the high saturation magnetization of Sm_2Co_{17} and the high magnetic hardness of $SmCo_5$ was ultimately reached by carefully controlling the kinetics of the precipitation reaction in a material of the approximate composition $SmCo_{7.7}$, in which some of the cobalt was replaced by iron and small amounts of copper and zirconium. High coercivities were obtained by applying a precipitation hardening treatment which resulted in the formation of bipyramidal particles of the 2-17 phase (rich in iron and cobalt) together with lamellae of a more Sm-rich phase. The high coercivity is due to domain wall pinning by the lamellae, leading to energy products as high as 260 kJm^{-3} as achieved by Mishra et al. [6] in 1981.

In 1984 two independent reports appeared describing the preparation and properties of a novel permanent magnet material obtained by combining Nd, Fe and B [7,8]. This material is based on the ternary intermetallic compound $Nd_2Fe_{14}B$, having a tetragonal structure. The importance of this novel compound stems from the fact that its main component is Fe, which is much less expensive than Co. In the second place, this compound has a crystal structure that gives rise to a crystal field interaction in which Fe as well as Nd produce a uniaxial anisotropy suitable for permanent magnet application. As a consequence one may profit from the ferromagnetic coupling between the Nd and Fe moments and the comparatively high Nd moment to reach a high value for the saturation magnetization.

The discovery of $Nd_2Fe_{14}B$ initiated a worldwide search for other novel materials. This lead to the discovery of several other novel ternary compounds with promising properties. These ternary compounds include tetragonal $R_2Fe_{14}C$, isomorphous with $R_2Fe_{14}B$, and the interstitially modified compounds $R_2Fe_{17}C_x$ [9] and $R_2Fe_{17}N_x$ [10]. An extensive number of other novel ternaries can be characterized by the formula composition $RFe_{12-x}T_x$ where T is a stabilising nonmagnetic component. The applicability of these compounds and their interstitial modifications as permanent magnet materials is still under investigation. Earlier comprehensible reviews on rare earth based permanent magnet materials can be found in Refs. [11-20]

2. BASIC CONCEPTS AND MODELS

2.1. INTRODUCTION

Permanent magnetic materials are characterized by a broad magnetic hysteresis loop and a concomitant high coercivity. The hysteresis loop of a given permanent magnet material is preferably measured on a long cylindrical sample in order to exclude demagnetizing effects. The second quadrants of the hysteresis loop of two permanent magnets are shown in Fig. 1.

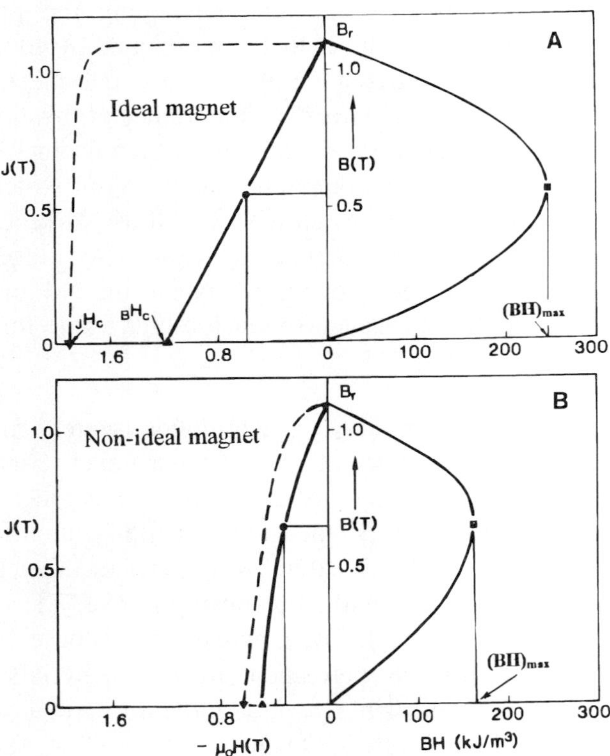

Fig. 1. *Demagnetizing behaviour of two magnets prepared from the same materials but manufactured under different conditions. In the left parts plots of flux density B (full lines) and magnetic polarization J (broken lines) are shown as a function of the demagnetizing field strength H. In the right parts the product BH (horizontal axis) for each point of the demagnetization curve of the magnet is plotted versus the corresponding B value (vertical axis). The maximum energy product (BH)_{max} is indicated on the B(H) curve for both types of magnets.*

Both magnets were made of the same starting material but were manufactured under optimum conditions (A) and under conditions far from the optimum (B). The remanence, B_r, determines the flux density that remains after removal of the magnetizing field and, hence, is a measure of the strength of the magnet. Note that the remanence values are equal for the two magnets considered, but that the magnetic polarisation J(T) falls off with increasing demagnetizing field much more strongly in case B than in case A. This has serious consequences for the coercivity, $_BH_c$, which is a measure of the magnet's resistance to demagnetizing fields.

The performance of a magnet is usually specified by its maximum energy product, that can be regarded as a figure of merit. This BH_{max} value and is defined as the maximum when forming the product of the flux density B and the corresponding opposing field H for each point of the B(H) curve in the second quadrant. In other words, a value for BH_{max} can only be given if measurements of the B(H) curve in the second quadrant have been made on a magnet body.

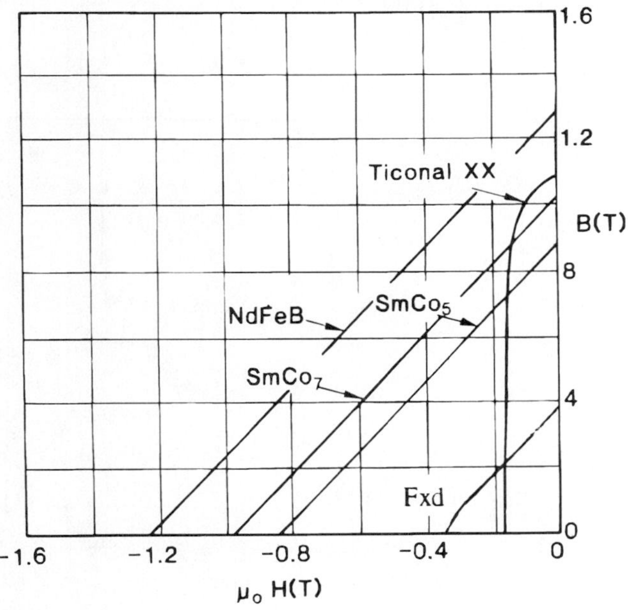

Fig. 2. *B versus H plots of several types of permanent magnets. Magnets of the type Sm(Co,Fe,Cu,Zr)$_{7.4}$ have been indicated by SmCo$_7$.*

Any statements about values of maximum energy products derived from loop measurements on powder samples are unrealistic and meaningless. This is also obvious from Fig. 1 showing that the maximum energy product for the manufacturing process used for magnet A is substantial larger than the maximum energy product in case B. The BH_{max} value should not be confused with the theoretical upper limit of the maximum energy product. The theoretical maximum energy product equals $1/4\ B_r^2$ and would have been obtained for an ideal magnet with a strictly linear B(H) curve up to at least $-H_{demag} = 1/2\ B_r$, as is the case for magnet A.

Besides the maximum energy product there are other criteria that can be used to specify the quality of a permanent magnet material. Of importance in many static applications is the magnitude of the intrinsic coercivity, $_jH_c$. This may be illustrated again in Fig. 1, which compares the field dependence of the magnetic polarization, J, and the flux density, B, of a magnet material. It follows from the relation $B = J + \mu_o H = \mu_o M + \mu_o H$, that $_jH_c$ and $_BH_c$ will not differ much when the former value is smaller than the remanence of the permanent magnet material, as is the case for the material in Fig. 1. In permanent magnet materials based on rare earth compounds, the intrinsic coercivity, $_jH_c$, can become much larger than the remanence, so that the B(H) curve can, in favourable cases, be linear in the entire second quadrant.

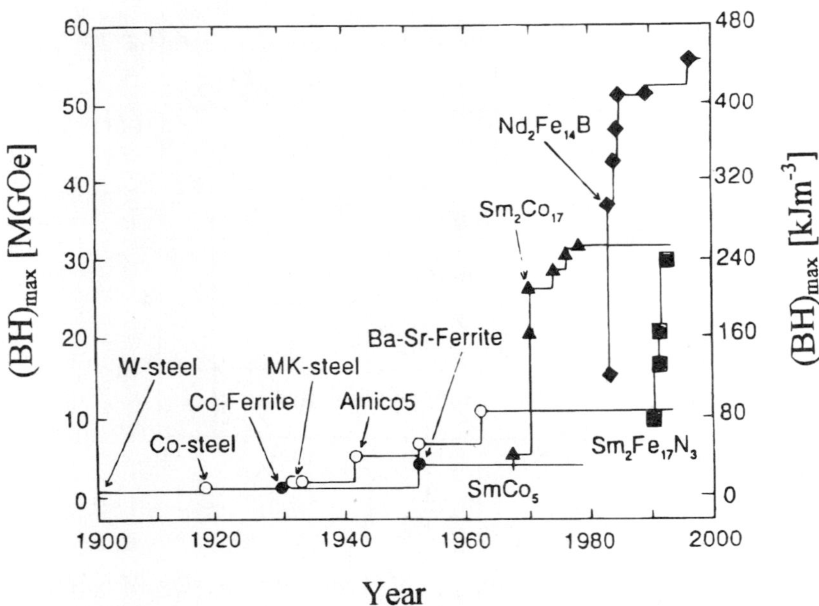

Fig. 3. *Survey of the development of permanent magnet materials. Permanent magnets based on rare earths have been indicated by full symbols.*

High values of $_jH_c$ can generally be obtained in magnet materials that have a high intrinsic magnetocrystalline anisotropy, as in rare earth compounds. In materials where the hard magnetic properties originate from shape anisotropy, as in alnico type materials, it is not possible to generate large coercivities. The B(H) curves of representative rare earth based magnets can be compared with the B(H) curve of Ticonal XX (alnico type) and some other magnet materials in Fig. 2. It is the presence of large coercivities, in particular that makes the rare earth based magnets suitable for applications requiring flat magnet shapes. The relative importance of the various types of permanent magnet materials can be derived from the diagram shown in Fig. 3.

Base Material	T_c		B_s		H_A		$(BH)^{th}_{max}$	
	[K]	[°C]	[kG]	[T]	[kOe]	[MAm^{-1}]	[MGOe]	[kJm^{-3}]
$Nd_2Fe_{14}B$	585	312	16.0	1.60	67	5.4	64.0	512
$Nd_2Fe_{14}C$	535	262	15.0	1.50	95	7.6	56.2	450
$Fe_3B : Nd$	785	512	16.0	1.60	--	--	64.0	512
$SmFe_{11}Ti$	585	312	11.6	1.16	92	7.4	33.5	268
$SmFe_{10}V_2$	610	337	11.0	1.10	60	4.8	30.	240
$SmFe_{10}Mo_2$	460	187	9.7	0.97	>50	>4.0	23.5	188
$Sm_2Fe_{17}C$	572	267	14.2	1.42	53	4.2	50.4	403
$Sm_2Fe_{17}N_{2.7}$	750	477	15.0	1.50	140	11.2	56.2	450
$SmCo_5$	1020	747	11.4	1.14	300	24	32.5	258
$Sm(Co, Fe, Cu)_7$	1100	827	13.	1.3	50	4.0	42.	336
$Nd_2Co_{14}B$	995	722	11.5	1.15	100	8.0	33.3	266

Table 1. *Magnetic characteristics of several RE based permanent magnet materials. The energy products (BH)$_{max}$ represent the theoretical upper limits. From Buschow [14].*

It can be verified from this figure that high performance permanent magnets are based on rare earth metals and that this type of permanent magnets has been developed in the last few decades. The main characteristics of some relevant rare earth intermetallics are listed in Table 1. The values of the theoretical maximum energy products listed may serve as a guide as to what results ultimately could be achieved with these materials under the most favourable circumstances. Practical values will be discussed in section 3.

2.2. MAGNETIC COUPLING AND ANISOTROPY

2.2.1. RIGOROUS TREATMENT

The two-sublattice mean field model is most suitable for describing the magnetic coupling between the magnetic moments in a lattice consisting of N_R rare earth atoms and N_T transistion metals atoms. If $B_{mol,R}$ and $B_{mol,T}$ represent the molecular fields experienced by the R moments (M_R) and T moments (M_T) one may write

$$B_{mol,R} = n_{RT} M_T + n_{RR} M_R \approx n_{RT} M_T \qquad (1)$$

$$B_{mol,T} = n_{RT} M_R + n_{TT} M_T \qquad (2)$$

where $M_T = N_T m_T$ and $M_R = N_R m_R$ represent the T and R sublattice moments, respectively. The quantities n_{RT}, n_{RR} and n_{TT} are the molecular field coefficients. the main interaction in the present class of materials is the T-intrasublattice interaction, followed in strenght by the R-T intersublattice interaction. It can be shown that the R-intrasublattice interaction is fairly weak. For this reason it is generally justified to neglegt it, as has been done in the right side of eq. (1).

For describing the magnetic structures as a function of temperature and applied field (B_0) one has also to take the magnetocrystalline anisotropy of the two sublattices into account.

It is frequently assumed that the main mechanism responsible for the magnetocrystalline anisotropy in solid 3d systems is the combined effect of spin-orbit interaction and partial quenching of the orbital angular momentum. The spin is coupled to the orbital motion via the spin-orbit coupling. Changes in spin direction will therefore be accompanied by directional changes in the orbital motion. The latter motion will, in turn, favour particular crystallographic directions dictated by electrostatic fields and overlapping wavefunctions associated with neighbouring atoms in the lattice. Modern electronic band structure calculation made by Nordström et al. [21] and Daalderop et al. [22,23] have shown, however, that this is an oversimplification. For details the reader is referred to the original papers. Here we will describe the T sublattice anisotropy by means of the phenomenological espression

$$E_{A,T} + K_{1,T}\sin^2\theta \qquad (3)$$

The total free energy expression for the T sublattice is then given by

$$E_T = K_{1,T} \sin^2\theta - M_T (B_{mol,T} + B_0) \qquad (4)$$

where θ is the angle between M_T and the c-axis.

The rare earth sublattice anisotropy is much better understood although it can also be fairly complex in character. It originates from the crystal-field induced single-ion contributions to the anisotropy. The latter can be derived from standard crystal field theory, described in more detail in the review of Hutchings [24]. The crystal field splitting of the (2J+1)-fold degenerate ground state multiplet levels and the corresponding wave funtions can be described in terms of the well known Stevens operators O_n^m

$$H_{\text{crystal field}} = N_T \Sigma B_n^m O_n^m \tag{5}$$

where the summation is over n and m. The quantities B_n^m represent the symmetry and strength of the crystal field surrounding the 4f electron system. The total Hamiltonian for a given rare earth site can be written as

$$H_R = \Sigma_{nm} B_n^m O_n^m - 2\Sigma_T J_{RT} S_R \cdot S_T - 2\Sigma_R J_{RR} S_R \cdot S_R \tag{6}$$

where J_{RT} and J_{RR} are the R-T and R-R magnetic coupling constants, respectively. After introducing the exchange field B_{ex}, commonly considered as acting on the spin moment, one has

$$H_R = \Sigma B_n^m O_n^m + 2S_R \cdot B_{ex,R} - m_R B_0 \tag{7}$$

when introducing the molecular field B_{mol} commonly considered as acting on the total moment ($<m_R> = - g\mu_B <J_Z>$ if z is the quantization axis) one has

$$H_R = \Sigma B_n^m O_n^m - m_R (B_{mol,R} + B_0) \tag{8}$$

where $B_{mol,R}$ is defined as in eq. (1). With $S_R = (g_R-1)J_R$, one finds

$$B_{mol,R} = B_{ex,R} - 2 (g_R - 1)/g_R = \gamma B_{ex,R} \tag{9}$$

By comparing eq. (6) with eq. (1) one easily derives

$$2\mu_B^2 N_T n_{RT} /\gamma = \Sigma_T J_{R\,T} \approx z_{RT} J_{RT} \tag{10}$$

where the last term is obtained when the R-T interaction is restricted to the first nearest T neighbours (z_{RT}) to a given R atom only.

It is good to recall that the Hamiltonian in eq. (8) applies to a single R ion. The total free energy of the R sublattice at a given temperature is obtained after diagonalization of H_R of eq. (8) and by using the resulting energy values to calculate the partition function Z_R. The free energy is subsequently obtained by means of the relation

$$F_R = k_B T \ln Z_R \tag{11}$$

The total energy of the two-sublattice system becomes

$$E_{tot} = E_T + \Sigma F_R + n_{RT} M_R M_T, \tag{12}$$

where the last term has been included to avoid double counting of the intrasublattice interaction terms.

It is now possible to determine the magnetic structure at any given temperature and applied field by solving the coupled eqs. (4) and (8) while minimizing the total energy of eq. (12). A common procedure is to minimize the energy for a given set of parameters B_n^m, n_{RT}, $K_{1,T}$ as a function of the direction of the sublattice magnetizations. The calculated values of the magnetization components in the direction of B_0 can subsequently be compared with experimental values available for different fields B_0 and at various temperatures. For more details regarding these procedures the reader is referred to papers published by Franse and Radwanski [25], Cadogan et al. [26], Yamada et al. [27] and Li and Coey [13].

The procedure described above lends itselfs well to a detailed description of the magnetization curves obtained at various temperatures and in various crystallographic directions on single crystals. From an analysis of such data it is possible to obtain values of B_n^m, n_{RT} and possibly n_{RR} for a given type of compounds, which then can be compared with the corresponding values of other types of compounds or with theoretical estimates of these quantities.

2.2.2. APPROXIMATE METHODS

The 3d sublattice anisotropy was introduced in the previous section by means of the relatively simple phenomenological relation, eq. (3). In practice it proves helpful to describe also the rare earth sublattice anisotropy by such a phenomenological expression, although in this case more terms in the expression for the anisotropy energy have to be considered. For measurements made on magnetically aligned powder particle of materials with a uniaxial crystal structure is it generally sufficient to use the folowing expression for the rare earth sublattice anisotropy:

$$E_{A,R} = K_{1,R} \sin^2\theta + K_{2,R}\sin^4\theta \tag{13}$$

where θ represents the angle between the c axis and the R sublattice magnetization. From the transformation properties of the Stevens operator equivalents the following relations between the anisotropy constants K_i and the crystal field parameters of eq. (5) may be obtained [28,29]:

$$K_{1,R} = - 3/2\ B_2^0 <O_2^0> - 5 B_4^0 <O_4^0> \qquad (14)$$

$$K_{2,R} = 35/8\ B_4^0 < O_4^0> \qquad (15)$$

A well known method to obtain experimental values of $K_1 = K_{1,T} + K_{1,R}$ is that described by Sucksmith and Thompson [30] and is based on the relation

$$2\ K_1 / J_s^2 + (4K_2/J_s^4)J_2 = H/J \qquad (16)$$

which holds for the magnetization curve of a single crystal or aligned powder particles in comparatively small fields applied perpendicular to the easy direction. When H/J is plotted versus J^2 the anisotropy constant K_1 in relation (16) may be derived from the horizontal intercept and the anisotropy constant K_2 from the slope of the straigth line.

It follows from eqs. (14) and (15) that the temperature dependence of the rare earth sublattice anisotropy constants depends on the thermal averages $< O_n^m >$. The expressions of O_n^m are polynomials of order n of J and J_Z (see the report of Hutchings [24]). For instance,

$$O_n^m = 3\ J_Z^2 - J(J+1) \qquad (17)$$

Since it can be shown that the expectation values of $< O_n^m >$ fall off with increasing temperature as $[M_R (T)/ M_R (O)]^{n(n+1)/2}$ this means that the higher the order of the anisotropy constant the lower its contribution at higher temperatures. This is the reason why for some classes of compounds, such as the $R_2Fe_{14}B$, $R_2Fe_{14}C$ and $R_2Co_{14}C$ series, only the K_1 term needs to be taken into consideration at room temperature, K_1 being sufficiently described now by the first term of eq. (14):

$$K_{1,R} = - 3/2\ B_2^0 <O_2^0> = - 3/2\ \alpha_J <r^2> A_2^0 < O_2^0> \qquad (18)$$

where α_J is the second order Stevens factor and $<r^2>$ the expectation value of the Hatree-Fock 4f radius.

Experimental information of the crystal field parameters A_2^0 is frequently obtained from rare earth Mössbauer spectroscopy. It can be shown that the following relation exists between A_2^0 and the principal component of the electric field gradient V_{zz} determined by rare earth Mössbauer spectroscopy [31-38]

$$A_2^0 \ [\ Ka_0^{-2} \ ] = - \ \omega V_{zz} \ [\ 10^{21} \ Vm^{-2} \ ], \tag{19}$$

The proportionality constant ω in this relation will be further discussed below. A discussion as to the validity of Eq. (19) can be found in Refs. [39,40].

In the case of complicated crystal field interactions ^{155}Gd Mössbauer spectroscopy is most useful because it samples exclusively sign and magnitude of the second order crystal field parameter, whatever the sign and magnitude of the higher order crystal field parameters.

Computational results of V_{zz} and A_2^0 for a group of compounds related to permanent magnet materials and obtained by various authors from electronic band structure calculations can be compared with experimental results in Table 2. In general, the agreement between experimental values of V_{zz} and theory can be said to be satisfactory. It is interesting to note that different types of electronic structure calculations do not lead to the same answer. Surprisingly, however, the corresponding calculated values of the ratio $\omega = - A_2^0 / V_{zz}$ are rather close to each other. Inspection of the data listed in the table furthermore shows that the calculated w ratios are also close to the experimental ω ratios. This can be taken as support for the applicability of relation (19), although there is no real fundamental basis for its validity. A more detailed discussion of this matter is given in the report of Buschow et al. [40].

Since the 4f charge cloud is spherical symmetric the field gradient V_{zz} present at the Gd nuclei as determined from ^{155}Gd Mössbauer spectroscopy can be used directly in eq. (19). In other types of rare earth Mössbauer spectroscopy, involving for instance ^{161}Dy, ^{166}Er, or ^{169}Tm the field gradient at the rare earth nuclei consists primarily of the contribution $V_{zz,4f}$ due to the asymmetric 4f charge cloud. This contribution has to be subtracted from the total experimental value of V_{zz} to yield the contribution due to the aspherical charge distribution outside the 4f shell. The latter is commonly indicated as lattice contribution $V_{zz,lat}$ and is used to calculate A_2^0 via equation (19). Details of such procedures can be found in publications of Friedt et al. [32], Gubbens et al. [51], Fruchart et al. [52], Vasquez and Sanchez [53] and Buschow et al. [54].

Compound	V_{ZZ}[cal]	V_{ZZ}[exp]	A_2^0[cal]	A_2^0 [exp]	w[cal]	w[exp]
$GdNi_5$	+16.2[41]	+9.7[41]		-450[42]		46
$GdCo_5$	+14.0[41] +18.3[23] +10.8[44]	+10.1[41]	-691[43] -972[23] -584[44]	-400[25]	49 [41,43] 53 [23] 54 [44]	40
Gd_2Co_{17}	+6.8[45]	+ 4.8[45]	-169[46]			35
Gd_2Fe_{17}	+7.5[46] +4.8[48] +6.4[49]	+ 4.3[47]	-302[43] -302[49]		40 [46,43] 47[49]	
$Gd_2Fe_{17}N_x$	+10.5[49] +11.0[48]	+12.6[45]	-475[49]		45[49]	
$Gd_2Fe_{17}C_x$	+13.8[49] +14.9[48]	+15[14]	-613[49]		44[49]	
$Gd_2Co_{17}N_x$	+14.8[45]	+14.9[45]				
$Gd_2Fe_{14}B(f)$	-8.8[46]	-7.75[50]	+371[46]	+300[14]	42[46]	39
$Gd_2Fe_{14}B(g)$	-7.7[46]	-7.67[50]	+381[46]	+300[14]	49[46]	39

Table 2. *Experimental values of electric field gradients and second order crystal field parameters [exp] in comparison with results derived from band structure calculations [cal]. The data are given in the following units: V_{zz} [10^{21} V/m^2], A_2^0 [Ka_0^{-2}]. For definition of the ratio ω, see main text. (*Component of the electric field gradient in the c direction)*

In Fig.4 examples are shown of how the development of the electric field gradient and the concomitant values of A_2^0 and K_1 can be followed easily by means of ^{155}Gd Mössbauer spectroscopy. As will be discussed in more detail in section 3.4, carbon and nitrogen atoms can occupy interstitial holes available in R_2Fe_{17} compounds. The maximum hole filling in the corresponding interstitial solid solutions $R_2Fe_{17}M_x$ is reached for x=3. As shown in Fig. 5, the M=N,C atoms occupy the interstitial holes surrounding the R atoms and hence strongly modify the electric field gradient at this site. The results shown in Fig. 4 illustrate how V_{zz}, A_2^0 and K_1 (eqs. 19 and 18) strongly increase with interstitial hole filling, explaining why the interstitially modified materials eventually become materials suitable for permanent magnet applications.

Another method frequently used to obtain information on the crystal fields is based on the so-called point charge approximation. In this approximation it is assumed that the electrostatic field acting on the 4f electron system is produced by point charges Z_j of the ligand atoms located at a distance R_j (x_j,y_j,z_j) from the rare earth ion considered. As mentioned, the leading term determining the magnetocrystalline anisotropy is A_2^0 in most of the compounds considered in this report. The corresponding expression in the point charge approximation is

$$A_2^0 = -1/4\,|\,e\,|\,\Sigma_j\,Z_j\,(3z_j^2 - R_j^2)/R_j^5 \qquad\qquad (20)$$

Similar expressions can be derived for the other A_n^m terms [24]. Since the distances R_j (x_j,y_j,z_j) are known fairly accurately in a given structure, it is in principle possible to calculate values of A_n^m by using expressions of the type (20) if values for the ligand charges Z_j were available. Owing to screening effects and hybridization effects the application of the point charge approximation with fixed charges for the various types of ligand elements remains doubtful.

Even more serious is the fact that not the charges of the ligand elements but rather the asphericities of the on-site valence electron charge clouds of the rare earths play the most important role in crystal field effects. Band structure calculations made on several rare earth based intermetallic compounds have largely confirmed the importance of the valence electrons of the rare earths in determining sign and magnitude of the second order crystal field parameter. Examples of such calculations can be found in reports of Zhong and Ching [56], Coehoorn et al. [39], Coehoorn and Buschow [46], Daalderop et al. [23], and Steinbeck et al. [57].

In permanent magnet materials one is mainly interested in the anisotropy at and above room temperature. From eq. (18) it can be derived that the rare earth sublattice anisotropy for a given value of A_2^0 depends on temperature through the thermal averages $<O_2^0>$ of the operator equivalents. Because of the smallness of JRR one might expect that the R sublattice is not able in itself to produce sufficiently high values of $< O_2^0>$ above cryogenic temperatures. For this reason high values of the intersublattice coupling constant J_{RT} are not only desirable but are essential for the application of R-3d intermetallics as permanent magnet materials. This has led to a systematic investigation of the dependence of the intersublattice coupling constant J_{RT} on the composition and nature of the R and T components [58].

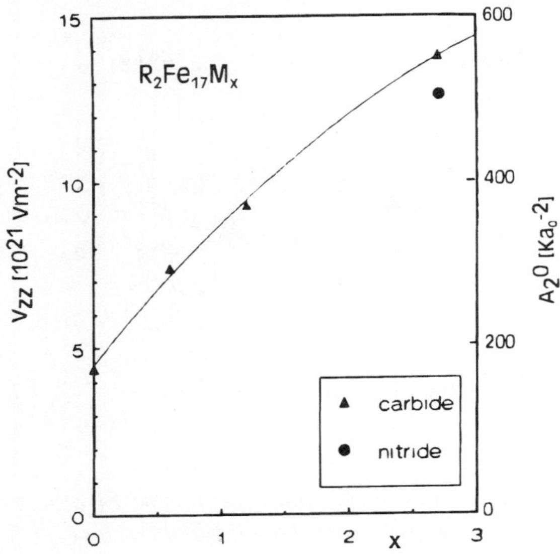

Fig. 4. *Effect of interstitial hole filling on the electric field gradient and second order crystal field parameter in R_2Fe_{17} compounds. The V_{zz} values were taken from the report of Dirken et al. [47). The V_{zz} value $Gd_2Fe_{17}C_{2.7}$ was taken to be equal to that of the corresponding Mn compound reported by Dirken et al. [55]. The value $\omega=40$ was used to transform V_{zz} into A_2^0.*

There are two experimental methods by means of which the coupling constants J_{RT} can be derived relatively easily from experiments, even from magnetic measurements made on powder samples. Below a brief outline of these two generally applicable simple methods will be given.

In the first method n_{RT} is derived from magnetic measurements made on powder particles in high fields at low temperatures [59]. The powder particles have to be sufficiently small in size so that they can be regarded as an assembly of small single crystals, able to rotate freely and orient their magnetization according to the external field. In many types of 4f-3d compounds, the anisotropy of the 4f sublattice exceeds that of the 3d sublattice by at least one order of magnitude at 4.2 K. By minimizing the free energy expression Verhoef et al. [59] showed that under such circumstances the low-temperature magnetization curve comprises two regions. Below $B_{1,crit}$ there is a strictly antiparallel alignment between the (heavy) rare earth moments and the 3d moments: $M = |M_R - M_T|$.

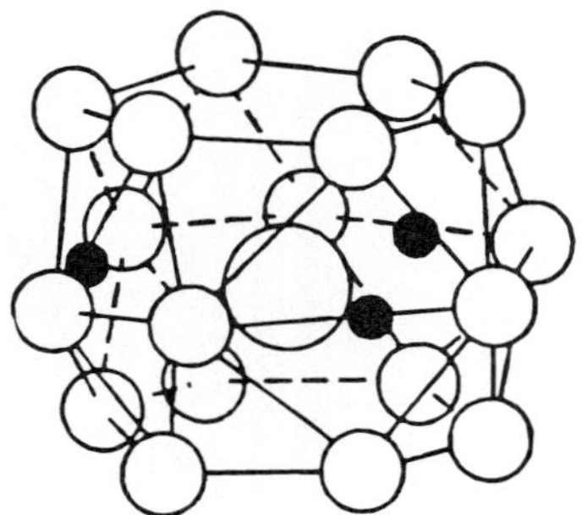

Fig. 5. *Coordination of the rare earth atoms (center) by Fe atoms (small open circles) and interstitial carbon or nitrogen atoms (small filled circles)*

For sufficiently high values of the applied field, $B_o > B_{2,crit}$, the R and 3d moments are parallel: $M = M_R + M_T$. In the intermediate field range, $B_{1,crit} < B_o < B_{2,crit}$, there exists a canted moment configuration, the R and T sublattice moments bending towards each other with increasing B_o. In this region the field dependence of the total moment is given by

$$M = B_o / n_{RT} \qquad (21)$$

The slope of the $M(B_o)$ curve in the linear intermediate regime lends itself therefore relatively straigthforwardly to the determination of experimental values of n_{RT}, from which the constant J_{RT} can be obtained via eq. (10). A prerequisite for this method is that the two sublattice magnetisation do not differ much in absolute value. The first critical field

$$B_{1,crit} = n_{RT} \, | M_R - M_T | \qquad (22)$$

is sufficiently low in this case so that the linear region given by eq. (21) falls into the experimentally available field range.

In general it was found that J_{RT} is almost temperature independent This means that reliable values of n_{RT} can also be derived in comparatively low fields for compounds having a compensation point in their temperature dependence of the magnetisation. When measuring the field dependence of M at the latter temperature one has $B_{1,crit} = 0$, and eq. (20) applies already for low fields starting from the origin.

In the second method use is made of the high-temperature approximation of the mean field model. At high temperatures it is reasonable to neglect the R-R interaction relative to the R-T and, a fortiori, relative to the T-T interaction. The standard mean field expression for the Curie temperature then reads as

$$J^2_{RT} = 9k^2 T_{c,R} (T_{c,R} - T_{c,o})/4z_{RT}z_{TR}S_T(S_T+1)G \qquad (23)$$

where the z_{TT}, z_{RT} and z_{TR} represent coordination numbers in the crystal structure considered and G is the de Gennes factor $G = (g_R-1)^2 J_R(J_R+1)$. For the determination of J_{RT} one has to substitute in eq. (23) the values of the Curie temperatures of compounds in which $M_R \neq 0$ ($T_c = T_{c,R}$) in conjunction with Curie temperatures of compounds (R = Y, La, Lu) in which $M_R = 0$ ($T_c = T_{c,o}$).

It should be borne in mind that problems in applying eq. (23) may arise when the T-T interaction varies across the rare earth series, because the experimentally easily accessible values of $T_{c,o}$ and S_{Fe} of compounds in which R = La, Lu or Y, when substituted in eq. (23) may lead to mutually inconsistent values of J_{RT}.

Electronic band-structure calculations dealing with magnetic intersublattice coupling have been performed by Brooks et al. [60]; Liebs et al. [61.62]; Hummler and Fähnle [44,63] Uebele et al. [48]. Brooks et al. [60] advocate that the key role in the R-T coupling is played by the 5d electrons of the R component. The spin-up and spin-down rare-earth 5d band states mix with the transition-metal 3d band states but do so to a different degree, depending on the degree of exchange splitting between the spin-up and spin-down 3d bands. The 5d(R)-3d(T) mixing leads to a larger occupation of the 5d spin-down band than the spin-up band, and the 5d(R) moment is therefore antiparallel to the 3d(T) moment. This entails an antiferromagnetic coupling between the 3d and 4f spins, since one has ferromagnetic intra-atomic exchange interaction between the 4f-spin moment and the 5d-spin density. Brooks et al. [60] calculated the effective exchange field in a perturbative way while Liebs et al. [61,62]; and Liu et al. [58] used a different approach, avoiding any specific model regarding the 4f-3d exchange mechanism. Both latter groups of authors agree with Brooks et al. that it is the rare-earth-valence electrons (mainly 5d) that mediate the 4f-3d exchange interaction, but emphasize the self-consistent nature of the calculations of the interaction constant. A comparison of computational results with various types of experimental results, including data obtained by inelastic neutron scattering, is presented in more detail in the review of Buschow [20].

2.3. COERCIVITY

When applying a steadily increasing magnetic field in a direction opposite to the easy magnetization direction of a perfect single crystal of a ferromagnetic compound, one would expect that all atomic moments reverse their orientation by a process of collinear rotation whenever the applied field becomes equal in size to the anisotropy field, H_A (uniform rotation). Already in 1948 Stoner and Wohlfarth [64] showed that for spheroid particles, in which the major axis coincides with the easy axis as determined by the magnetocrystalline anisotropy, the coercivity is given by

$$H_c = 2 K_1 / J_s - (N_{||} - N_\perp)J_s , \qquad (24)$$

where K_1 is the first-order anisotropy constant, $J_s = \mu_o M_s$ is the saturation polarisation, $N_{||}$ and $N_\perp$ are the demagnetizing factors in the easy and hard direction, respectively.

In practice, however, coercivities are substantially smaller than expected on the basis of eq. (24). In fact, the majority of permanent magnet materials gives rise to magnetization reversal at field strenghts that appear to be only a small fraction (approximately 15%) of the value of $H_A = 2 K_1 / \mu_o M_s$. The reason for the comparatively easy magnetization reversal is the occurrence of magnetic domain structures. Magnetic particles of sufficiently large size will generally not be uniformly magnetized. Instead, they will be composed of magnetic domains that are mutually separated by domain walls or Bloch walls. The magnetizations in adjacent domains point in opposite directions in order to keep the magnetostatic energy low. The magnetization in the wall between two domains gradually changes from one preferred magnetization direction to the other. The thickness of the wall is determined by the relative strength of the anisotropy energy and the exchange energy. The former tends to reduce the wall thickness, the latter tends to increase the thickness.

Domain walls, and the concomitant reversed domains, can be nucleated relatively easily near lattice imperfections, where the local values of the exchange and anisotropy fields are sufficiently reduced from the values in the bulk of the material to make a local magnetization reversal possible. This wall nucleation at defects may take place spontaneously although it is facilitated by an externally applied negative magnetic field.

The field required for wall nucleation, commonly referred to as the nucleation field, H_N, is an important parameter for describing the coercivity, H_c. Non-uniform processes are usually encountered, in which the magnetization does not take place by uniform rotation, but by wall nucleation and propagation. These processes dominate in materials with high magnetocrystalline anisotropy. By analogy with eq. (24) an empirical relation of the type [65-67]

$$H_c = \alpha \psi^{min} \, \alpha_K \, K/J_s - N_{eff} J_s \qquad (25)$$

has proven quite useful for describing the nucleation field, H_N, and the corresponding coercivity, H_c. The quantities a and N_{eff} are microstructural parameters. The first of these, $\alpha \psi^{min}$, accounts for particle misalignment whereas α_K accounts for the effect of inhomogeneities at the grain boundary and exchange decoupling. The parameter N_{eff} represents the influence of local effective demagnetizing fields.

The coercivity is mainly determined by difficulties in reverse domain nucleation in permanent magnets of the so-called nucleation-type. Once a wall has been formed adjacent to a nucleated reverse domain it can move comparatively easily further into the grain. For obtaining high coercivities the wall motion must remain restricted to the affected grain only. Hence wall motion must be impeded by grain boundaries, because otherwise a single nucleated wall would lead to magnetization reversal of the entire magnet. Wall pinning at grain boundaries is therefore considered to be a prerequisite for nucleation-type magnets.

A different mechanism is operative in the so-called pinning type magnets, where wall nucleation can proceed comparatively easily but where the domain walls cannot travel freely throughout the whole grain. Magnetic inhomogeneities present in the grains may act as pinning centers, impeding wall motion and preventing further magnetization reversal. Wall displacement, other than small amounts of wall bending, can occur only when the force exerted on the wall becomes sufficiently strong. This is the case when the strength of the external field exceeds the pinning field, H_p, which then determines the coercivity, i.e. $H_C = H_p$. More details regarding the coercivity mechanisms described above can be found in reviews published by Givord et al. [68,69] and Kronmüller et al. [66].

Alloys of the type $Sm(Co,Fe,Cu,Zr)_{7.4}$ are examples of permanent magnet materials that are pinning-controlled. The alloy consists of a single phase at high temperatures. Subsequent heat treatment of the material at lower temperatures leads to the formation of a finely divided precipitate that is able to pin the Bloch walls and produce a high coercivity. The permanent magnet materials $Nd_2Fe_{14}B$ and $SmCo_5$ are examples where the coercivity is nucleation-controlled, which has important consequences for the manufacturing of these magnets. Any impurity phases able to nucleate reverse domains and walls have to be avoided in the microstructure of these nucleation-controlled permanent magnets. An ideal microstructure for nucleation-controlled magnets consists of fine grains of the main magnetic phase, each grain being magnetically isolated from the other grains by a thin layer of intergranular material. The latter material need not necessarily be single phase but should not contain magnetic phases. When the nonmagnetic layer is fairly homogeneous the value of α in eq. (25) can become rather large. Small amounts of a magnetic material in the intergranular region can act as wall nucleation centers and destroy the coercivity. Of course, also the grains of the main phase have to be free of any inclusions or precipitates of soft magnetic phases. For the manufacture of permanent magnets it is therefore desirable to know the phase relationships of the corresponding systems. For Nd-Fe-B and Sm-Co type magnets, these phase relationships will be briefly discussed in section 3.

In the last few years computer simulations of the magnetisation reversal behaviour of realistic microstructures have been used for studying microstructural effets. Such calculations can be found in reports of Schrefl and Fidler [70], Uesaka et al. [71], Schrefl et al. [72-74], Feutrill et al. [75], Schrefl and Fidler [76,77], and Kronmüller et al. [78]. These computer simulations are based on finite element calculations and deal with quasi-real homogeneous or inhomogeneous distributions of phases and grain diameters, such as can be observed by electron microscopy.

This method entails minimisation of the total magnetic Gibbs free energy in an applied field, and includes the exchange energy, the magnetocrystalline anisotropy energy, the stray field energy and the magnetostatic energy of the magnetic polarization in the external field. Particular types of microstructures can be generated by using a random or a biassed distribution of nuclei for grain growth as starting point. Grain growth can subsequently be programmed to proceed either isotropically or along preferred directions. The generated grains are then triangulated for the finite element calculations, along with programmed distributions of hard, soft and nonmagnetic regions. A considerable advantage is that the coercivity of such microstructures can be calculated numerically for various degrees of average misorientations of the grains. A more detailed discussion of these methods is presented in Ref. [20].

3. RARE EARTH BASED PERMANENT MAGNET MARERIALS

3.1. NdFeB TYPE MAGNETS

3.1.1. PHYSICAL PROPERTIES AND PHASE RELATIONSHIPS

Structural and magnetic characteristics for compounds of the $R_2Fe_{14}B$ type have been listed in the top part of Table 3. Although permanent magnet manufacturing is virtually restricted to $Nd_2Fe_{14}B$, several of the other rare earth metals are used as additives in order to optimise the permanent magnet properties. The data listed in the table may serve as a guide to obtain information in which direction magnetization, Curie temperature or anisotropy are modified when substituting one or more of the other rare earth components for Nd. In the same way the data listed in the bottom part of the table for a selected number of $R_2Co_{14}B$ may serve to estimate the effects of substituting Co for Fe.

Material	T_C [K]	T_{SR} [K]	M_S (4.2K) $[\mu_B/fu]$	J_S (300K) [T]	μ_oH_A (300K) [T]	A_2^0 $[Ka_0^{-2}]$	A_4^0 $[Ka_0^{-4}]$
			$R_2Fe_{14}B$				
R=La	530	--	30.6	1.38	2	--	--
R=Ce	422	--	29.4	1.17	3.0	--	--
R=Pr	569	--	37.0	1.56	8.7	220	5.3
R=Nd	568	135	37.7	1.60	6.7	304	- 14
R=Sm	620	--	33.3	1.52	--	--	--
R=Gd	659	--	17.7	0.89	2.5	--	--
R=Tb	620	--	13.2	0.70	22.0	303	- 13
R=Dy	598	--	11.3	0.71	15.0	294	- 13
R=Ho	573	58	11.2	0.81	7.5	312	- 12
R=Er	551	320	12.9	0.90	--	299	- 14
R=Tm	549	311	18.1	1.15	--	303	- 12
R=Yb	524	115	--	--	--	152	- 7
R=Lu	534	--	28.5	1.17	2.6	--	--
R=Y	571	--	31.4	1.41	2.0	--	--
			$R_2Fe_{14}C$				
R=Pr	513	--	34.8	1.27	14.8	--	--
R=Nd	532	120	31.4	1.41	9.5	--	--
R=Gd	630	--	18.1	0.73	3.5	--	--
R=Lu	459	--	27.2	1.16	3.1	--	--
			$R_2Co_{14}B$				
R=La	955	--	19.9	0.74	--	--	--
R=Pr	995	664	24.8	0.83	10	--	--
R=Nd	1007	36,54	25,5	0.89	4.5	--	--
R=Sm	1029	--	18.1	0.99	--	--	--
R=Gd	1050	--	5.3	0.71	--	--	--
R=Tb	1035	794	2.2	0.91	--	--	--
R=Y	1015		19.5	0.85	--	--	--

Table 3. *Magnetic characteristics obtained on single crystals of $R_2Fe_{14}B$ compounds and on powders of $R_2Fe_{14}C$ and $R_2Co_{14}B$ compounds. The data were taken from the report of Buschow [14].*

Many reports dealing with investigations of the Nd-Fe-B phase diagram have appeared in the literature since the discovery of $Nd_2Fe_{14}B$ in 1984. Results obtained in several of these phase diagram investigations were reviewed by Buschow[14], and further references of reports dealing with phase diagram information useful for the manufacture of Nd-Fe-B type magnets can be found in a later review of permanent magnet materials [20].

In this section only the most important features of the Nd-Fe-B phase diagram will be discussed. It may be seen from the isothermal section at 900 °C reproduced in Fig. 6a that there are three ternary compounds, viz. $Nd_2Fe_{14}B$ (ϕ), $Nd_{1+e}Fe_4B_4$ (η) and $Nd_5Fe_2B_6$ (ρ). The diagram shown has to be regarded as an oversimplification, because binary as well as ternary compounds are indicated as line compounds. For the sake of clarity any ranges of homogeneity have been left out of consideration. The tie line shown in Fig. 6a between the h phase and Fe is no longer present at temperatures above 900 °C. According to the results of Schneider et al. [79] a tie line between the phases f and Fe_2B is found instead. Schneider et al. also showed that the type of phase transformations observed when Fe-rich liquids are cooled depends strongly on the degree of superheating of the alloys during melting of the constituent elements.

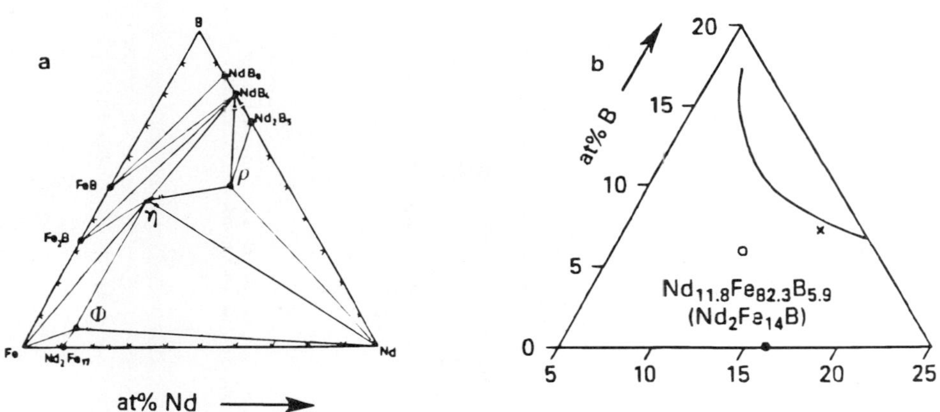

Fig. 6. *(a).Ternary section of the Nd-Fe-B system at 900°C (Nd-poor alloys) and 600°C (Nd-rich alloys). The three ternary phases are Nd₂Fe₁₄B (ϕ), Nd₁₊ₑFe₄B₄ (η) and Nd₅ Fe₂B ₆ (ρ). (b). Fe rich corner of the Nd-Fe-B system. For concentrations above the full curve primary crystallzation of Fe can be avoided.*

For temperatures below 900 °C one has to include in Fig. 6a also the occurrence of a further binary compound, Nd_5Fe_{17}. The latter compound was first observed by Schneider et al. [80]. It is a stable phase in the Nd-Fe system formed peritectically at 780 °C [81]. Its crystal structure was determined by Moreau et al. [82]. This structure is hexagonal and comprises 7 crystallographic inequivalent Nd sites and 14 inequivalent Fe sites.

The various types of phase transformations that may occur when dealing with cooling Fe rich melts from the liquid state can better be assessed by means of a different way of representing the phase relationships in the Fe rich corner of the Nd-Fe-B system. To that end we show in Fig. 7 a vertical section passing through the composition $Nd_2Fe_{14}B$. It follows from the results shown in Fig. 7 that, when a molten alloy of the composition $Nd_2Fe_{14}B$ is cooled infinitely slowly, so that equilibrium is reached at each intervening temperature, one would observe the following phases. On cooling from high temperature, the liquidus is traversed for the composition $Nd_2Fe_{14}B$ at about 1280 °C. Further cooling then leads to the crystallization of crystals of pure Fe from the melt, their number and magnitude increasing steadily until, at about 1180 °C, the peritectic melting point of the $Nd_2Fe_{14}B$ phase is reached. Below the peritectic melting point the Fe crystals and the remaining less Fe-rich liquid phase are no longer equilibrium phases and have to transform into $Nd_2Fe_{14}B$.

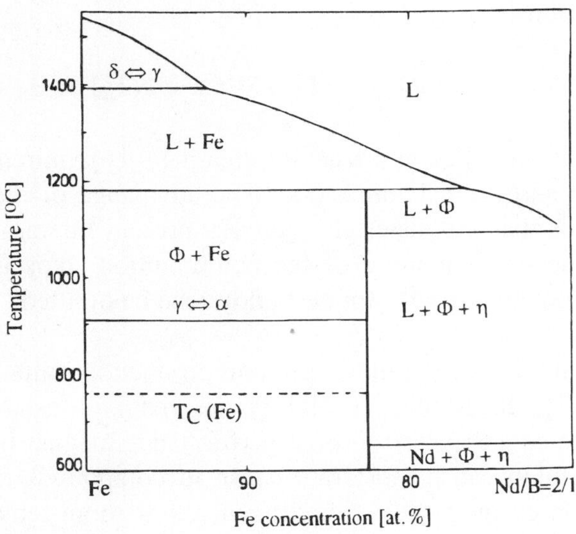

Fig. 7. *Vertical section in the Nd-Fe-B phase diagram for compositions with a Nd/B ratio of 2 (from Schneider [79]).*

One has to bear in mind, that formation of the $Nd_2Fe_{14}B$ phase will start on the outside of the Fe crystals. The latter crystals together with the liquid phase will ultimately become completely consumed. For sufficiently slow cooling rates only the $Nd_2Fe_{14}B$ phase would be left, this being the phase stable at room temperature. The $Nd_2Fe_{14}B$ phase formed around the Fe crystals below 1180°C represents, however, a diffusion barrier for the reaction of the Fe in the interior with the less Fe-rich liquid. This means that the peritectic formation of $Nd_2Fe_{14}B$ will be incomplete in a normal casting process. Residues of primary Fe crystals will be left in the $Nd_2Fe_{14}B$ grains, which are in turn surrounded by a relatively Nd-rich liquid that has solidified at the eutectic temperature at about 630 °C. Is has been mentioned already in the previous section that the presence of the primary Fe crystals is harmful for the attainment of a sufficiently high coercivity, because the Fe can serve as nucleation centres for reverse domains. This is one of the reasons that permanent magnet alloys are generally chosen in a concentration range where the crystallization of primary Fe is avoided. These concentrations are found above the solid curve drawn in the diagram of Fig. 6b.

3.1.2. SURVEY OF MAIN MANUFACTURING ROUTES OF NdFeB TYPE

The starting alloys are usually prepared by induction melting of the constituent elements in a protective atmosphere. Alternatively the starting material can be prepared by calciothermic reduction [83] performed at 1200 °C and involving the following reaction:

$$8\ Nd_2O_3 + 4\ B_2O_3 + 5\ Fe_2O_3 + 66\ Fe + 51\ Ca \Rightarrow Nd_{16}Fe_{76}B_8 + 51\ CaO$$

After completion of the reaction a leaching treatment is required to remove the unreacted calcium and the calcium oxide. The advantage of the calciothermic reduction process is that the material is already present in small particle form. For this reason one or even more of the comminution steps that have to be applied in some production routes for cast alloys can be omitted.

A diagram representation of the most common production routes for cast alloys is displayed in Fig. 8. Routes A and B are the well-established powder metallurgical treatments that lead to high-performance magnet bodies. Route B is particularly useful because most rare earth 3d compounds can readily and reversibly absorb large quantities of hydrogen gas at room temperature and at moderate pressures. A survey of the effect of hydrogen in hard magnetic materials can be found in the report of Fruchart et al. [84].

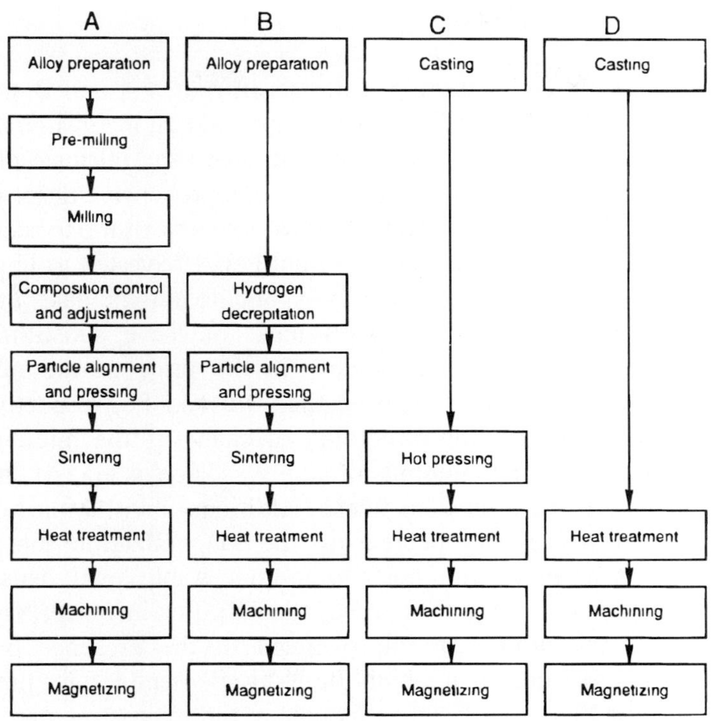

Fig. 8. *Survey of main manufacturing routes in the production of different types of rare-earth permanent magnets.*

A strong volume increase accompanies the absorption of hydrogen gas. This leads to pulverization of the material which, when used in combination with attritor milling or jet milling, is most effective in producing fine Nd-Fe-B particles. In this so-called HD (hydrogen decrepitation) technique the cast, coarse grained, $Nd_{15}Fe_{77}B_8$-type alloy is readily broken up into a relatively fine powder by exposure to hydrogen at around 1 bar pressure and room temperature. A total of about 0.4 weight percent hydrogen is absorbed, first by the neodymium rich phase, and then by the $Nd_2Fe_{14}B$ matrix phase. After milling, the powder is aligned, pressed, and vacuum sintered. This last process removes all the hydrogen from the sample, at the same time forming a fully dense body with a fine grain size [85]. It follows from the results of detailed studies made by several investigators (see, for instance, references given in the review of Buschow [20]) that the desorption of hydrogen during the sintering process occurs in three stages. First, hydrogen is released by the $Nd_2Fe_{14}B$ matrix phase at about 200°C and subsequently by the neodymium-rich intergranular phase in two stages at about 250 and 600°C.

The alloys generally used as starting materials in routes A and B show a large variety in compositions. Here it should be mentioned that a considerable body of experimental studies has been carried out to improve the intrinsic properties of $Nd_2Fe_{14}B$ by different types of substitutions. A comprehensive description of the results of these studies is given in Ref. [12]. Double substitutions such as Dy for part of the Nd and Co for part of the Fe in $Nd_2Fe_{14}B$ have led to improvements in temperature coefficient of the coercivity and to an increase in the Curie temperature. Drawbacks associated with most of these substitutions are a diminished magnetization and an increase in the price of the raw material because Dy and Co are more expensive than Nd and Fe, respectively. Other investigations have explored the possibility of changing the microstructure of the sintered magnets by small amounts of additives such as Ga, Al, and Nb. The changes in microstructure due to these additions involve an increase in coercivity without necessarily decreasing the magnetization too much and without increasing the production costs to an undesirable level. This is true for double additives, such as Co and V or Mo [86-88]. For instance, Sagawa et al. [86] showed that the prominent improvement of the presence of V is the generation of a microstructure in which the $NdFe_4B_4$ is no longer present in the intergranular region. The beneficial influence on the microstructure stems from the uniform distribution of small $V_{3-x}Fe_xB_2$ precipitates that can suppress excessive growth of the $Nd_2Fe_{14}B$ grains during liquid phase sintering, growth which would have decreased the coercivity. This makes it possible to raise the sintering temperature with a concomitant more complete consumation of Nd, Fe, and $NdFe_4B_4$ in the liquid phase to form $Nd_2Fe_{14}B$. The ultimate result are sintered magnets with improved coercivities and corrosion resistance.

It has proven possible to avoid the fairly expensive powder metallurgical route, as indicated by route C in Fig. 8 [89-91]. The magnet bodies obtained by this process are, however, isotropic and have energy products considerably below those of the magnets obtained via routes A and B. A further simplification is obtained in route D. This method did not lead to useful values of the energy product in the case of NdFeB alloys, but proved very suitable when applied to PrFeB alloys containing additives [92]. Energy products up to 145 kJ/m^3 were reported.

Rapid solidification processing such as melt-spinning and atomization can be used to prepare alloys having a particularly fine-grained microstructure. Because this manufacturing process is a less common one it has not been included in Fig. 8. Coarse powders prepared from melt-spun ribbons (Magnequench powder) are commercially available. Hot pressing of the powder at about 750°C leads to full densification without impairing the coercivity as a result of excessive grain growth, as described by Lee [93] and Lee et al. [94]. Typical values of the energy product are 100 kJ/m^3. But magnets with energy products over 120 kJ/m^3 have also been reported. When the hot pressing is followed by die-upsetting, textured magnet bodies are obtained with their easy magnetization direction parallel to the direction of press. The increase in remanence so obtained is responsible for the fairly large energy products.

In contradistinction to NdFeB powder obtained by milling or grinding of normal cast alloys, powder obtained from melt spun alloys is coercive and can therefore be used for bonded magnets. The powder is incorporated into a polymer binder by injection or compression moulding. The energy product is necessarily small for the corresponding magnets because of the reduced volume fraction of the magnetic material. The main advantage of the bonded magnets is the possibility of giving the magnet bodies the final required shape with close dimensional tolerances. An alternative way of obtaining coercive powders is described at the end of the following section.

3.1.3. LATER DEVELOPMENTS

3.1.3.1. Sintered magnets

In the preceding section it was mentioned already briefly that the manufacture of sintered magnets based on $Nd_2Fe_{14}B$ involves starting alloys that are fairly off-stoichiometric. Generally lower Fe concentrations are used, the reason for this being the avoidance of primary α-Fe that is magnetically soft and hence hampers the generation of large coercivities. Furthermore, lower Fe concentrations lead to the presence of sufficient amounts of Nd-rich intergranular material. The latter magnetically isolates the $Nd_2Fe_{14}B$ grains, which promotes large coercivities. In order to have a high H_C it is necessary, in general, that the particles remain small and magnetically well isolated after sintering. In principle, it is not required that the total amount of intergranular material be as high as, for instance, in the standard $Nd_{15}Fe_{77}B_8$ alloy. After casting the latter alloy consists of large $Nd_2Fe_{14}B$ grains with fairly large amounts of intergranular material between the grains. This inhomogeneous distribution of the intergranular materials is undesirable and becomes only partially improved after the milling treatment. As a result, the distribution of the Nd rich intergranular material remains inhomogeneous also in the sintered magnets. For this reason more intergranular material is required for magnetic isolation of the $Nd_2Fe_{14}B$ grains than would actually be needed. This means that it is possible to manufacture sintered magnets with improved values of the coercivity, remanence, and energy product by better control of the microstructure, where an important role is played by the amount and distribution of the Nd rich intergranular phase.

Because of the realization that the coercivity is more than sufficient in many room temperature applications, several later investigations have focused on improving primarily remanence and energy product. Superficially, the steps to be taken to reach this goal look fairly easy, because all that has to be done is to improve the alignment of the grains during compacting of the powder in the presence of a field, and/or to increase the main phase at the cost of the intergranular phase. It is mainly the highly peritectic nature of the main phase, discussed in section 3.1.1, that presents problems when one tries to increase the relative amount of the main phase. These problems can only partially be solved by changing the phase relationships via additives.

A curve defining the Fe rich concentration region in which primary Fe forms is shown n Fig. 6b [79]. In practice, concentrations below the curve but close to it may be used because particle size reduction of the alloy may bring the small primary Fe crystals in direct contact with the Nd-rich liquid during sintering, causing the disappearance of these crystals during the latter treatment. When using alloys of even higher Fe concentration, i.e. concentrations close to the formula composition $Nd_2Fe_{14}B$, the primary Fe crystals are too large and generally will not disappear during sintering. The α-Fe in such alloys can be removed before sintering by a heat treatment of the cast ingots. Because the Nd-rich intergranular material is virtually absent, such annealed alloys of nearly stoichiometric composition have the additional advantage that oxygen pick-up during the powder metallurgical treatment can be kept low. This greatly facilitates the handling of the corresponding powders, which is a significant advantage during magnet manufacturing. However, sufficient intergranular for magnetic isolation has to be present in the ultimate sintered magnets and this has to be provided in the form of a second powder of Nd rich composition, to be blended with the powder of the annealed alloy before sintering.

A particularly interesting manufacturing route based on such principles was reported by Otsuka and Otsuki [95]. These authors were able to obtain sintered magnets with fairly high values for the maximum energy product, $(BH)_{max} \approx 400$ kJ/ m^3 and remanence, $B_r \approx 1.52$ T by using concentrations very close to the stoichiometric composition for the main alloy. The corresponding powder (92 Vol%) was mixed before sintering with a second powder (8 Vol%) of higher Nd content (about 60 wt% Nd). The latter powder served as a sintering aid and was obtained from alloys prepared by continuous splat cooling and present in amorphous or microcrystalline form. The coercivity of the magnets mentioned is comparatively modest ($\mu_0 {}_jH_c \approx 1$ T) but the authors reported that higher values for $_jH_c$ were obtained when using splat cooled sintering aids containing several types of additives. Also it proved possible to choose these additives in a way as to enhance the corrosion resistance of the intergranular phase.

The use of a sintering aid in the form of a second powder prepared by splat cooling is expensive, as is the homogenization treatment of the main alloy. A possible way for cost reduction is time reduction of the homogenisation treatment. This was reached by Ahmed et al. [96] by using tiny amounts of additives such as Nb or Zr that can suppress the occurrence of primary Fe crystals in the main melt. A further cost reduction can be obtained by using a sintering aid that is not amorphous but can be prepared by standard casting. This second powder needs not necessarily be composed of the same elements contained in the main phase. The only requirement is that after sintering the second powder has reacted with the main phase and/or the original intergranular phase to lead to a homogeneous distribution of the ultimate intergranular material and that the latter is able to magnetically isolate the main phase particles. The feasibility of using various types of sintering aids in the form of second powders of various simple compositions has been described already by Gandehari [97], Sasaki et al. [98], Hamamura et al. [99] and Kaneko [100].

A more detailed investigation as to the effect of additives in the form of a second powder has been made by de Groot et al. [101]. These authors used powder of the standard alloy, $Nd_{14.2}Fe_{78.6}B_{7.2}$ that was thoroughly mixed small amounts of a second powder consisting of DyGa before sintering (90 min. at temperatures between 1090 and 1120°C). When the samples were given a post-sintering treatment at temperatures above 640°C, de Groot et al. report a large reversible drop in coercivity which is accompanied by decomposition of the δ-phase ($Nd_6Fe_{14-x}Ga_x$). The latter phase, together with Nd_3Ga_2, were shown to be present in the intergranular material. De Groot et al. [101] analysed their data by means of eq. (25) for which one expects a linear relationship when H_c/ J_s is plotted versus $\alpha\psi^{min} H_A / J_s$. The corresponding plots, reproduced in Fig. 9, show that the relatively low coercivity in magnet annealed above 640°C is caused by a decreased value of α_K rather than by a decreased demagnetizing factor N_{eff}. This was taken as evidence that the coercivity decrease is not caused by any soft magnetic phase formed during post-sintering above 640°C. By means of electron probe micro analysis (EPMA) it was furthermore shown that Dy is only present in those parts of the $Nd_2Fe_{14}B$ grains that had formed during the grain growth accompanying the sintering step. Similar Dy distributions were also reported by Ghandehari [102] using Dy_2O_3 as second powder and Velicescu et al. [103] using Dy_2Co_3.

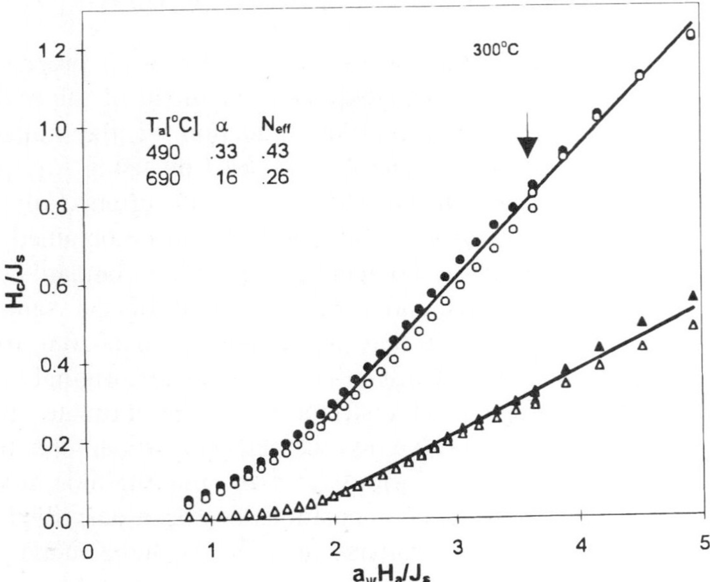

Fig. 9. *Coercivity versus anisotropy field obtained at various temperatures on sintered two-powder magnets magnets based basically on $Nd_2Fe_{14}B$ powder and DyGa powder. The upper and lower curve were obtained after post-sintering at 490 °C and 690°C, respectively. From de Groot et al. [101].*

The effect of adding second powders of low melting elements (Al, Mg) or high melting elements (W, Mo) on remanence and coercivity were studied by Yan et al. [104]. Sintered magnets produced from a mixture of hydrogen decrepitated NdFeB powder and variable amounts of Cu powder were investigated by Ragg and Harris [105] and shown to increase the coercivity without detrimental effect on the remanence. Generally, it can be stated that a beneficial aspect of the two-powder method is the following. The excess Nd originally present in the eutectic phase can be removed because it combines with the non-rare earth component of the second powder to form a new type of nonmagnetic intergranular material. The latter material leads to better magnetic isolation of the grains and to enhanced corrosion resistance.

A somewhat different strategy to improve the performance of sintered magnets has been focused on the high-temperature properties of sintered NdFeB magnets. The NdFeB magnets have a very undesirably large temperature coefficient of H_c. so that they cannot be used in electromotive applications at operating temperatures around 200°C. The reason for the large temperature coefficient of H_c is the equally large temperature dependence of the anisotropy field, which in turn is associated with the strong decrease of $< O_n^m >$ in eq. (18) when approaching the Curie temperature. In most of the high-temperature applications extremely high remanence are not required, which opens the possibility to increase H_c via a concomitant increase of the anisotropy by substituting Dy and/or Tb for part of the Nd, as may easily seen when inspecting the HA values listed in Table 2. Substitution of the latter elements implies a decrease of the saturation magnetization (see Table 2) because the heavy rare earth moments couple antiparallel with the Fe moments. The main effect of these substitutions is to increase the absolute value of the anisotropy field and coercivity while there is only little improvement of the temperature coefficients. However, the room temperature values of the coercivity are that large, that a sufficiently high value remains at the high use temperature of these magnets. Needless to say that also in the high-temperature NdFeB alloys, a strict control of the microstructure is required to optimize the coercivity so that the amount of Dy (Tb) additives can be kept as low as possible. Also the nature of the ultimate intergranular material is of paramount importance in these types of magnets because it has to be highly corrosion-resistant for the high-temperature applications envisaged.

3.1.3.2. Hot-formed magnets

The powder metallurgical route is avoided in the production of hot-formed magnets (see route C in Fig. 8). Melt-spun NdFeB alloys have been used as starting materials to produce different grades of magnets, the best results being obtained by die-upsetting as described already by Brewer [106] and Saito et al. [107]. Later Mishra et al. [108] have been able to improve the properties of die-upset magnets by using melt spun NdFeB ribbons containing small quantities of Co, Ga, and C. The energy product was reported to be close to 400 kJ/ m^3 with $B_r = 1.48$ T and $\mu_{o\,j}H_c = 1.48$ T. The latter result was achieved mainly through controlled grain growth during densification and die-upsetting, and by keeping the amount of intergranular phase low. Similar results were obtained by Shinoda et al. [109]. These authors also used Co and Ga additives and showed that optimum results with BH_{max} close to 400 kJ/ m^3 were reached for the compositi-on $Nd_{13.2}Fe_{72.5}Co_{7.5}Ga_{0.8}B_6$.

The manufacture of other types of hot formed magnets having lower performance, but simultaneously requiring considerably lower energy expenditure has also been reported. As described by Perrier de la Bâthie [110], hot extrusion of ingots imbedded in protective iron sheets leads directly to cylindrically-shaped magnets for rotating machines with either tangential or radial magnetisation. The energy product is lower than for sintered magnets but one may benefit from the lower costs. Useful results were obtained also wit Cu containing RFeB ingot alloys as described by Nozières et al. [90] and Akioka et al. [111]. Ingots of the composition $Pr_{20}Fe_{74}Cu_2B_4$ were used by Kwon et al. [112] to obtain magnets by an upset forging process in which the deformation was carried out in approximately 20 secs. The same authors attributed the magnetic alignment occurring during the upset forging to grain boundary gliding of the plate-like grains ($B_r = 1.05$ T, $\mu_{0\,j}H_c = 1.2$ T). Also in these less expensive magnet brands there is the possibility to profit from near-net shape manufacturing.

3.1.3.3. Bonded magnets

It was mentioned already that fine powders prepared from cast or annealed NdFeB alloys by grinding do not possess sufficiently high coercivities for application of these powders in resin bonded magnets. However, coercive powders can be obtained by grinding sintered or hot-formed magnets. Also alloys obtained by rapid solidification processing such as melt-spinning and atomisation can lead to coercive materials, as described by Sellars et al. [113]. A fairly economical route to coercive powders consists of the HDDR process as described by Takeshita et al. [114], Harris [115], Takeshita and Nakayama [116] and Nakayama and Takeshita [117]. This process consists essentially of four steps, hydrogenation of $Nd_2Fe_{14}B$ at low temperatures, decomposition of $Nd_2Fe_{14}BH_x$ into $NdH_2+\delta+$ Fe + Fe_2B, desorption of H_2 gas from $NdH_2+\delta$, and, finally, recombination of Nd + Fe + Fe_2B into $Nd_2Fe_{14}B$. Because the formation of $Nd_2Fe_{14}B$ grains in the last step is a solid state reaction, it proceeds at a much lower rate than during normal solidification from the melt. This has as a consequence that the average grain size remains small and leads to sufficiently large coercivities. The HDDR powders can be used to manufacture isotropic and anisotropic bonded magnets, as described by Nakayama et al. [118]. The energy product BH_{max} of an anisotropic bonded magnet based on Zr and Ga doped $Nd_2(Fe_2Co)_{14}B$ has been reported to reach values up to 143 kJ/ m^3 with $B_r = 0.89$ T and $\mu_{0\,j}H_c = 1.37$ T [116]. More details of the HDDR process and the concomitant phenomena are discussed in the review of Buschow [20], and in reports of Yartis et al. [119] and Liesert et al. [120]. Wide-spread applications of various types of bonded magnets will be given in section 7.

3.2. TERNARY CARBIDES WITH $Nd_2Fe_{14}B$ STRUCTURE

The magnetic properties of the $R_2Fe_{14}C$ compounds are almost the same as those of the corresponding borides. As can be derived from the data shown in the middle part of Table 2, the Curie temperatures and the saturation magnetizations are slightly lower whereas the crystalline anisotropies are slightly higher.

Although ternary carbides of the type $R_2Fe_{14}C$ are stable at room temperature for all the rare-earth elements, there are obvious difficulties in their formation for the first few R members when applying normal casting. This difficulty results from a solid-state transformation occurring at higher temperatures where the $R_2Fe_{14}C$ compounds decompose into compounds of the type $R_2Fe_{17}C_x$. The corresponding transformation temperature (T_t) is fairly low for the compounds with R=La, Ce and Pr. This hampers the formation of the $R_2Fe_{14}C$ phase during annealing after casting, because reaction rates have become too low for the formation of the $R_2Fe_{14}C$ phase from the peritectic reaction products at the annealing temperature, which necessarily has to be chosen below T_t. It is possible to shift the transformation temperatures in upward direction to some extent by substitutions.

The interest in these materials is primarily due to the presence of the mentioned solid-state transformation, because it offers the possibility of generating small-particle microstructures by controlled transformation of the high-temperature phase into the low-temperature $R_2Fe_{14}C$ phase in cast alloys. Alternatively, one could also say that nucleation and growth of the 2:14:1 phase occurs near the melting temperature in the Nd-Fe-B system and therefore leads to a large-grained microstructure. But in Nd-Fe-C the 2:14:1 phase nucleates and grows only in the solid state below T_t. Hence the microstructure can be more easily controlled and modelled into one resembling sintered magnets [121]. The so-called ingot magnets are magnetically isotropic. The advantage of these materials is their simple production route, no powder metallurgy or melt spinning being required. Attempts to prepare $Nd_2(Fe,Co)_{14}C$ type permanent magnets by mechanical alloying of powders of the elements (including various types of additives) were reported by Sui et al. [122].

3.3. SmCo$_5$ AND Sm$_2$Co$_{17}$ TYPE MAGNETS

3.3.1. SmCo$_5$ TYPE MAGNETS

Hexagonal compounds having the $CaCu_5$ type structure are formed for all common rare earth elements, albeit the stoichiometric 1:5 composition is only found for compounds of the lighter rare elements. Starting from R=Tb the composition becomes more Co rich, tending to RCo_6 for the heaviest rare earth elements. Simultaneously the eutectoid decomposition temperature of these phases shifts from about 600 °C for $SmCo_5$ to about 1200 °C for $ErCo_6$ [123]. The magnetic properties of the $CaCu_5$ type compounds have been described in several reviews [25, 124]. From these data it can be derived that only $SmCo_5$ qualifies as a starting material for permanent magnets.

The eutectoid decomposition of $SmCo_5$ involves phase separation into Sm_2Co_7 and Sm_2Co_{17}. Since the required long-range diffusion of atoms is very slow below 600 °C, this decomposition is very difficult to observe and plays no important role in the processing routes of permanent magnets. For this reason, the main manufacturing route of permanent magnets based on $SmCo_5$ is not much different from that described already in section 3.1.2 for $Nd_2Fe_{14}B$, and is based on liquid phase sintering. Cold pressed powder compacts of $SmCo_5$ are sintered to high density in an atmosphere of high purity inert gas slightly above 1100°C until about 95% of the theoretical density is reached. It is also possible to perform the sintering at lower temperatures when hydrogen gas is used instead of the inert gas. In the latter case it is necessary to apply a further heat treatment in which the absorbed hydrogen gas is removed from the sintered body by vacuum pumping. More details of the manufacturing process are given in the reviews of Kumar [125] and Strnat [11]. A model for calculating the best chemical composition when maximising coercivity and remanence has been presented by de Campos et al. [126]

Permanent magnets based on $SmCo_5$ are more widely used than those based on Sm_2Co_{17} to be discussed in the next section.. The attainable energy products are close to 180 kJ/ m^3 and the coercivities can reach 2.4 MA/m. Another favourable property of these magnets is their high Curie temperature, which is close to 750°C. The temperature coefficient of the coercivity is significantly lower than that of $Nd_2Fe_{14}B$ based magnets, but it is still too high for several high-temperature applications of these magnets. Because the coercivity almost vanishes near 475°C, the upper use temperature is generally restricted to below 250°C. During the years, the properties of $SmCo_5$ type magnets have been optimized by several types of modifications, in which other elements were substituted for part of the Sm and Co. The most prominent of these modifications have been discussed in several reviews [11,20,127]. A comparison of the properties of $SmCo_5$ type magnets with NdFeB type magnets will be made in section 3.3.3.

3.3.2. Sm_2Co_{17} TYPE MAGNETS

Permanent magnets in which the Sm_2Co_{17} phase is the main constituent owe their favourable properties to the presence of a homogeneity range of this phase. This homogeneity range extends mainly to more Sm rich concentrations, i.e. in the direction of the $SmCo_5$ phase [128]. The composition is chosen slightly Co deficient ($SmCo_7$). A single-phase solid solution is obtained after a homogenization treatment of $SmCo_7$ slightly below 1200°C. After cooling and an ageing treatment at about 800°C a fine precipitate of the magnetically hard phase $SmCo_5$ develops and act as pinning centres for domain walls (bulk hardening process). Magnets of the 2:17 type in commercial production now have a composition close to $Sm(Co,Fe,Cu,M)_{7.4}$. They contain some iron to increase B_r, copper to permit the magnetic precipitation hardening, and small amounts of other elements (M = Zr, Hf, Ti or a mixture of these) to generate microstructures needed for the precipitation hardening. The nature of the pinning sites in such microstructures has been studied by Wong et al. [129] using high resolution Lorentz microscopy. The rare earth component is usually Sm, but in some products this is combined with Ce, Pr or Nd owing to the relative scarceness of Sm. A schematic representation of the sequence of heat treatments is given in Fig. 10.

A comprehensible discussion of isothermal sections of the Sm-Co-Cu phase diagram, illustrating the reduction of the homogeneity range of the 2:17 phase when going to lower temperatures, has been given in the review of Strnat [11]. This reduction of the homogeneity range and the concomitant phase transformations are the reason that typical microstructures are obtained after a low-temperature heat treatment.

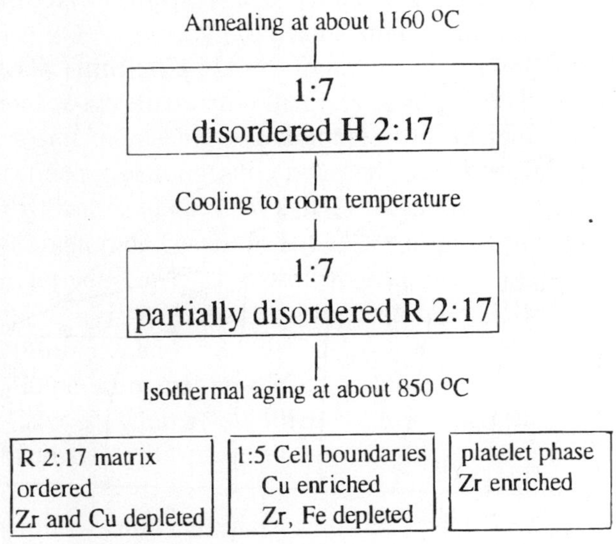

Fig. 10. *Heat treatments and the corresponding phase transformations involved in the manufacturing of $Sm(Co,Fe,Cu,M)_{7.5}$ type permanent magnets.*

The microstructures of $Sm(Co,Fe,Cu,M)_{7.5}$ type permanent magnets consist of a fairly well-developed network of very small cells of a matrix phase, within the much larger grains, which are separated and often completely surrounded by a thin boundary phase of 1:5 stoichiometry and $CaCu_5$ structure.

After a heat-treatment of the ingots leading to their optimum magnetic properties, the cells have linear dimensions of 100-200 nm and the cell walls are typically 5-20 nm thick. The rhombohedral (R) modification of the 2:17 structure is found in the cell interior where it appears in heavily twinned form, with the twin boundaries in the basal plane. Electron microscopic investigations reveal also other, very thin, layers that are parallel to the basal plane and run across many cells and cell boundaries. These are the fingerprints of a third phase, the so-called "platelet phase" or "z-phase" that contains most of the M element (Zr, Hf orTi). It has been suggested by Maury et al. [130] that these platelets have a crystal structure related to that of $SmCo_3$. All three phases are crystallographically fully coherent in good magnets. This is possible because simple relations exists between the lattice parameters of these phases.

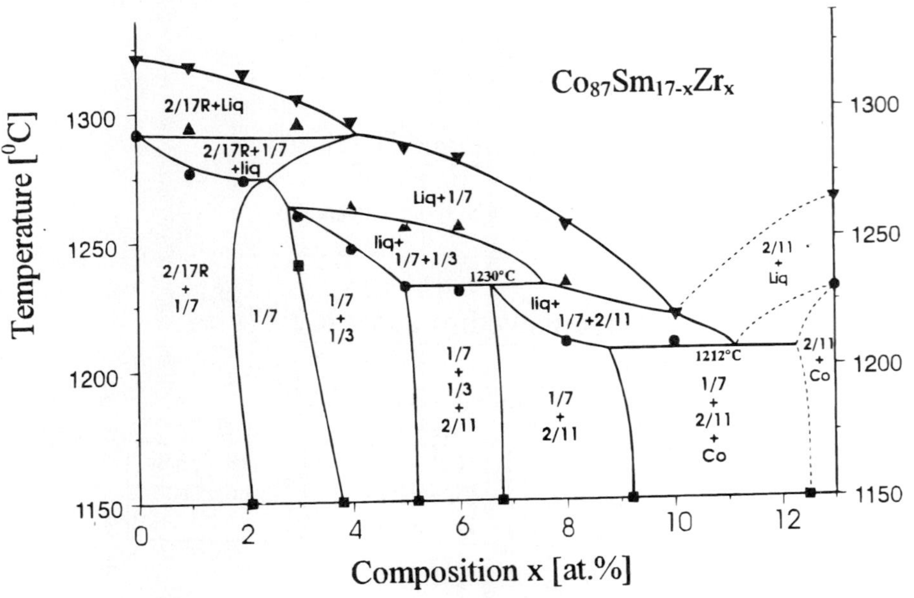

Fig. 11. *Phase relationships in the ternary Sm-Co-Zr system in the concentration region of interest for permanent magnet alloys. From Lefevre et al. [131]*

The phase relationships in the ternary Sm-Co-Zr system have been investigated in detail by Lefevre et al. [131]. An important detail of their study is reproduced in Fig. 11. Isoplethic sections for the three Co concentrations indicated by the horizontal dotted lines in the figure can be found in the original paper of Lefevre et al. One of these, pertaining to 87 a.% Co has been reproduced in Fig. 12.

It has been discussed already by Ray and Liu [127] and Maury et al. [130] that the high coercivity of the $Sm(Co,Fe,Cu,Zr)_x$ magnets depends strongly on the development of a fine-scale cellular microstructure containing the three phases mentioned above. This cellular microstructure of the three coherent phases develops from a single-phase metastable precursor alloy obtained after high-temperature annealing at about 1160°C. It has been assumed for many years that the crystal structure of the precursor alloy is rhombohedral and that there is considerable disordering associated with its off-stoichiometric composition. Cataldo et al. [132] and Lefevre et al. [131,133] showed, however, that the homogeneous single phase precursor alloy obtained after annealing at about 1160 °C is not of a disordered rhombohedral 2:17 type, but of the $TbCu_7$ structure type [134]. This can also be derived from the phase relationships shown in Figs. 11 and 12, where the solubility of Zr in the rhombohedral 2:17 is seen to be a very limited one. It is reasonable to assume that the Fe and Cu atoms in $Sm(Co,Fe,Cu,Zr)_x$ are located in one or more of the crystallographic Cu sites in the $TbCu_7$ structure type.

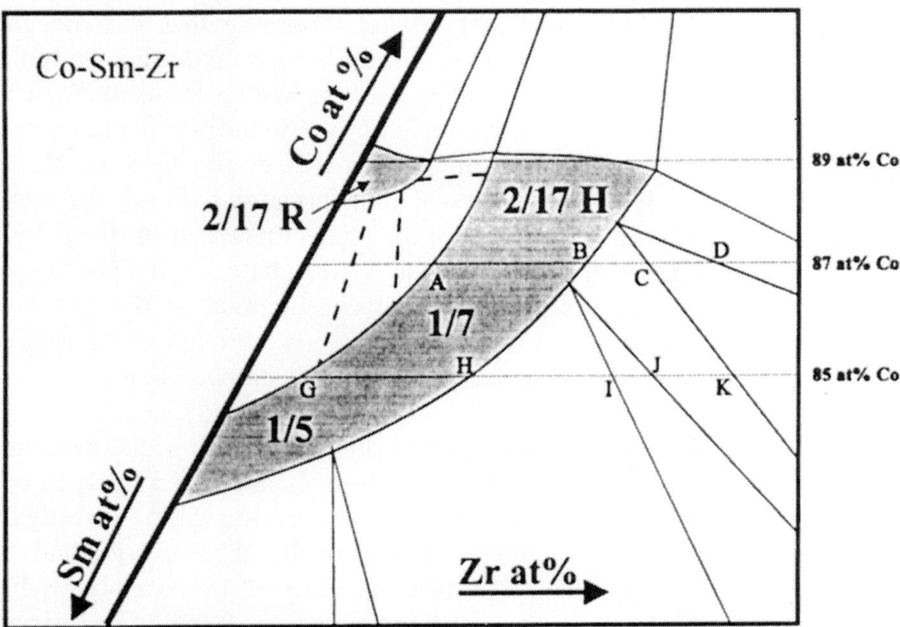

Fig. 12. *Isoplethic section of the Sm-Zr-Co phase diagram for 87 at.% Co. From Lefevre et al. [131]*

The situation is less clear regarding the location of the Zr atoms. It is well known that the hexagonal and rhombohedral 2:17 structures and the 1:7 structure can be derived from that of hexagonal RCo_5 ($CaCu_5$-type) by substitution of a pair of Co atoms (dumbbell pair) for some of the R atoms. The substitution of a single dumbbell pairs for Sm in $SmCo_5$ requires an energetically unfavourable expansion in the c-direction, which can be compensated in the fully ordered Sm_2Co_{17} only by a relaxation of the other types of Co atoms. It appears less likely therefore that the Zr atoms, being much larger than the 3d atoms, occupy one of the two dumbbell positions and leave a vacancy at the corresponding second dumbbell site, as proposed by Ray and Liu [127]. This view is refuted also by a structure determination made by Lefevre et al. [133] showing that Zr substitutes not into one of the dumb bell sites but randomly into the Sm sites.

The sequence of transformations by means of which the precursor alloy separates into the three phases present in the microstructure of the aged alloy is fairly complicated. Ray and Liu [127] propose that isothermal ageing leads to the formation of the ordered R 2:17 phase, which process is initiated by the precipitation of the Zr-rich platelet phase. After the stoichiometric R 2:17 has formed, its supersaturation by Cu leads finally to the precipitation of the Cu rich 1:5 cell boundary phase. A somewhat different sequence of transformation has been advocated by Maury et al. [130] who propose that the concentration fluctuations present in the off-stoichiometric precursor alloy first lead to a rapid nucleation and growth of ordered R 2:17 grains during the isothermal ageing. This implies that there is a relative increase of Cu and Sm in the surrounding matrix tending towards the 1:5 composition and eventually forming the cell boundary phase. A less rapid series of transformations during the isothermal ageing is involved in the formation of the platelet phase. It proceeds by means of Zr depletion and further ordering of the R 2:17 phase, i.e. the platelet phase is formed from the Zr together with partial consumption of the cell boundary phase. This means that during this process there is a growth of the R 2:17 cells and formation of the platelets at the cost of the 1:5 cell boundaries.

In spite of the complexity of the transformations involved in the manufacturing of $Sm(Co,Fe,Cu,Zr)_x$ type magnets many attempts have been made to optimize costs and performance of these magnets. Liu and Ray [135] investigated the effect of increasing Fe concentration and found that there is a gradual increase of the remanence. However, the benefits derived from applying high Fe concentrations are limited because the coercivity was found to decrease for the highest Fe concentrations considered. Liu and Ray [136] also varied the nominal composition x of the alloys and found that the optimum $(BH)_{max}$ value is obtained for a composition fairly close to x=7.74.

The high Curie temperature of $Sm(Co,Fe,Cu,Zr)_x$ type magnets leads to a low temperature dependence of the remanence and an only modest temperature dependence of the coercivity, even in the temperature range above room temperature. It has been shown in detail by Liu et al. [137] that even though the intrinsic coercivity $_jH_c$ decreases in the whole temperature range up to 250°C, it has little effect on $_BH_c$ and the energy product. For this reason $Sm(Co,Fe,Cu,Zr)_x$ type magnets are most favourable for high-temperature applications. Kim [138] and Ma et al. [139] have investigated the possibility of designing high-temperature permanent magnets with operating temperatures up to more than 450°C by proper control of the chemical composition and microstructure of $Sm(Co,Fe,Cu,Zr)_x$ type magnets.

Cost reduction was the main purpose of investigating the substitution of Ce for part of the Sm in $Sm(Co,Fe,Cu,Zr)_x$ type magnets, because one may profit from the much lower price of Ce compared to Sm. The Ce substitution is, however, accompanied by a substantial loss in magnet performance . The reason for this is not only that Ce is tetravalent and has no magnetic moment. The tetravalency of Ce additionally lowers the 3d transition metal moments and the Curie temperature. Several authors have investigated the possibility of substituting other light rare earth elements for Sm in $Sm(Co,Fe,Cu,Zr)_x$ [137,140, 141].

Substitutions of Nd and Pr, which both have a substantially higher atomic moment than Sm, open the possibility to increase the remanence, although without a distinct price benefit. It can be derived, however, from Eq. 18 that this type of substitutions lower the overall rare earth sublattice anisotropy because the sign of α_J for Pr and Nd differs from that of Sm. Consequently, also the coercivity is expected to be lowered (Eq. 24). Liu et al. [137] found that owing to the initial high value of $_jH_c$ there is an effect on $_BH_c$, however, only for fairly high Nd(Pr) substitutions. In fact, this means that one may profit from the remanence enhancement up to about 30% substitution of Nd or Pr.

3.3.3. COMPARISON WITH NdFeB TYPE MAGNETS

The permanent magnet materials of the $Sm(Co,Fe,Cu,Zr)_x$ and $SmCo_5$ type are characterized by nearly linear demagnetization curves. This favourable property is retained up to their maximum use temperatures, which are fairly high. Such nearly linear demagnetization curves can be reached in magnet materials based on $Nd_2Fe_{14}B$ only by substituting appreciable amounts of Dy or Tb for Nd. The latter substitutions enormously increase the room temperature coercivity so that the coercivity remains sufficiently strong also at high temperatures not too far above room temperature. The linear behaviour in the second quadrant of the hysteresis loop in $Sm(Co,Fe,Cu,Zr)_x$ and $SmCo_5$ type materials allows to use these magnets at very low permeance values without significant flux losses after exposure to appreciably higher temperatures. Such low permeance conditions may arise when using very flat magnets.

• Inspection of the phase diagrams of the systems Nd-Fe-B and Sm-Co shows that the main magnetic phase $Nd_2Fe_{14}B$ is in thermal equilibrium with Nd metal or with a Nd-rich phase in case of more exotic starting compositions. In the Sm-Co system all phases in thermal equilibrium with $SmCo_5$ are Co rich, and this remains true also for more complex alloy compositions. This implies that corrosive intergranular material will generally be absent in magnets of the Sm-Co types. In addition, also the pure main magnetic phases $Sm(Co,Fe,Cu,Zr)_x$ and $SmCo_5$ show higher corrosion resistance than $Nd_2Fe_{14}B$. Consequently, both the 1:5 and 2:17 type Sm-Co magnets exhibit better corrosion resistance than NdFeB type magnets, a fortiori at higher use temperatures.

• The Sm-Co type magnets have lower mechanical strengths than NdFeB type magnets.

• The maximum energy products of 2:17 type Sm-Co magnets fall in the lower end of the sintered NdFeB range. Generally they are higher priced. But the price difference can become relatively unimportant for small magnets where the advantage of the lower material cost does not play a significant role.

• A distinct disadvantage of Sm-Co magnets is the use Co metal as the majority component. Cobalt is fairly costly compared to iron. The strategic nature of Co metal, for which the price can fluctuate considerably, is the major drawback of the Sm-Co based magnets.

3.4.   INTERSTITIALLY   MODIFIED R_2Fe_{17} AND $R_3(Fe,M)_{29}$ COMPOUNDS.

The low Curie temperatures and comparatively low magnetocrystalline anisotropies make the R_2Fe_{17} compounds less attractive for applications as permanent magnet materials. Considerable improvements with respect to Curie temperature and anisotropy have been reached, however, by forming interstitial solid solutions obtained by combining these materials with carbon or nitrogen. The corresponding ternary nitrides and carbides $R_2Fe_{17}C_x$ and $R_2Fe_{17}N_x$ are generally believed to be restricted to compositions in the range $0 \leq x \leq 3$. More details about the formation ranges and location of the interstitial atoms in the lattice can be found in the review of Fujii and Sun [18].

A similar improvement in magnetic properties can also be reached in compounds of the type $R_3(Fe,M)_{29}$ which were discovered more recently and in which M represents Ti, V, Mn, Cr or Mo [142-146].

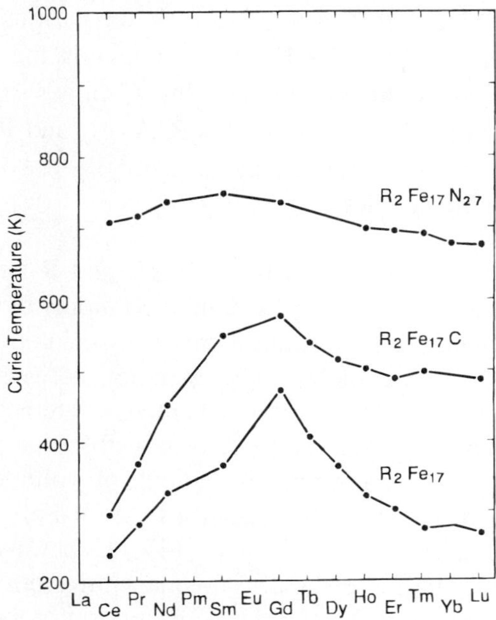

Fig. 13. *Effect on the Curie temperature of interstitial hole filling by carbon or nitrogen atoms in rare earth iron compounds of the type R_2Fe_{17}.*

It is relatively easy to prepare ternary carbides $R_2Fe_{17}C_x$ of comparatively modest C content ($x \leq 1.5$). This can be done by standard alloying techniques such as by arc melting of rare earths, Fe and C. For the preparation of materials with higher C concentrations one has to prepare the pure R_2Fe_{17} compounds first and subsequently expose the finely powdered material to hydrocarbon gasses at elevated temperatures. The ternary nitrides $R_2Fe_{14}N_x$ can be prepared by heating powdered R_2Fe_{17} in a nitrogen or NH_3 atmosphere. For more details see the reviews of Li and Coey [13] and Fujii and Sun [18].

Examples of how the interstitial hole filling leads to strong enhancements of the Curie temperature. Several examples are displayed in Fig. 13. Generally it is found that the Curie temperature in $R_2Fe_{17}C_x$ increases with small carbon concentrations x more strongly than for larger x values. However, there is a linear behaviour when T_c is plotted versus the corresponding unit cell volumes

[14, 147]. In fact, the Curie temperature enhancement can be correlated with results of earlier investigations of the compounds R_2Fe_{17} which had shown that their Curie temperatures decrease strongly under applied pressure. Brouha et al. [148] combined results of d T_c /dP with results of compressibility measurements and determined the value of Γ = d ln T_c /d ln V for various types of rare earth transition metal intermetallics. For R_2Fe_{17} compounds the value Γ = 13 was obtained, which compares favourably with the Γ values corresponding to the straight lines in plots of Δ T_c versus ΔV for $R_2Fe_{17}C_x$ and $R_2Fe_{17}N_x$ [14].These results indicate that the T_c enhancement is primarily a volume effect, in the carbides as well as in the nitrides.

The Curie temperature enhancement in $R_2Fe_{17}C_x$ and $R_2Fe_{17}N_x$ compounds is often related to the occurrence of the Fe dumb bell atoms in the crystal structure of the R_2Fe_{17} compounds. The volume increase associated with the interstitial hole filling reveals the short interatomic separation between these atoms. The latter is thought to lead to antiferromagnetic interactions between the corresponding Fe moments and is held responsible for the too low Curie temperatures in R_2Fe_{17}. Extensive investigations of volume effects in several interstitial and substitutional solid solutions of R_2Fe_{17} compounds show, however, that this is an oversimplification [149]. Generally, it seems unlikely that short interatomic Fe-Fe distances favour antiferromagnetic interactions and/or low Curie temperatures [150,151]. A discussion of the Curie temperature enhancement in terms of the spin fluctuation theory of Mohn and Wohlfarth [152] has been given by Woods et al. [153].

The strong Curie temperature enhancement is accompanied by only a rather modest Fe moment enhancement, as can be derived from high-field measurements made at 4.2 K by Liu et al. [154]. One has to bear in mind, however, that the Curie temperature enhancement leads to quite a substantial increase in the magnetization at room temperature, which is of importance for permanent magnet applications. In fact, because the N and C up-take leads to strong enhancements of the Curie temperature for $Sm_2Fe_{17}Z_x$ in particular, these materials are currently considered as good candidates for permanent magnet applications.

A further prerequisite for the application of these materials as permanent magnets is the magnetocrystalline anisotropy. It was mentioned already in section 2.2 that the occupation of interstitial holes close to the rare earth atoms (see Fig.5) leads to substantial changes of the second order field gradient A_2^0, which mainly determines the rare earth sublattice anisotropy. Generally, it has been observed by various types of rare earth Mössbauer spectroscopy [35,45,47] that the second order crystal field parameter A_2^0 tends to more negative values with x. Using eq. (18), this means that K_1 tends to more positive values with increasing x for all compounds in which the R component has a positive value of α_J. For this reason, the tendency of the R sublattice to align the moments along the c-axis becomes stronger with increasing C content for R elements with $\alpha_J > 0$, i.e. for $Sm_2Fe_{17}C_x$, $Er_2Fe_{17}C_x$, $Tm_2Fe_{17}C_x$ and $Yb_2Fe_{17}C_x$. It appears that the rare earth sublattice anisotropy (M $\parallel$ c) is strong enough to overcompensate the Fe sublattice anisotropy (M $\perp$ c) in $Sm_2Fe_{17}C_x$ already at fairly low carbon concentrations. For example, the easy magnetization direction in $Sm_2Fe_{17}Co_{0.5}$ is parallel to the c-axis at all temperatures below T_c [147] contrasting the behaviour of Sm_2Fe_{17} where the easy magnetization direction is perpendicular to the c-axis at all temperatures. Wang and Hadjipanayis [155] reported that a concentration as low as x = 0.1 is sufficient to give rise to an easy magnetization direction parallel to the c-axis at low temperatures. A more detailed discussion of the coercivity mechanisms in nitrogenated compounds has been given by Chin et al. [156]

By means of thermodynamic arguments [14] it can be shown that binary nitrides RN and Fe metal will be formed when the charging process during nitrogenation or (carbonization) is performed at too high a temperature or when the time during which the R_2Fe_{17} powder is heated under nitrogen gas or hydrocarbon gasses is too long. Liu et al. [154] found from a combination of magnetic measurements and X-ray diffraction that this degradation reaction starts at about 750 K. Owing to an increasing reaction rate the degradation process steadily progresses upon further heating and apparently reaches completion around 900 K, as reported by Katter et al. [157]. It is therefore generally not possible to process magnetic materials based on $R_2Fe_{17}N_x$ at temperatures above 900 K. In other words, the ternary carbides and nitrides can not be applied in the form of sintered magnets, as has been described for materials like $Nd_2Fe_{14}B$, $SmCo_5$ and Sm_2Co_{17} in sections 3.2 and 3.3. Sugimoto et al. [158,159] reported that the decomposition temperature can be shifted to some extent in upward direction by additives. But this is still far below temperatures where dense sintered magnets could be manufactured.

In general, only nitrides and carbides of Sm_2Fe_{17} qualify as starting materials for permanent magnets, because only these materials have remanences and coercivities high enough for permanent magnet applications. In order to generate coercivities of sufficient magnitude in these materials it is a prerequisite that the grain sizes be very small. This opens the possibility to five major routes for the preparation of coercive powders of interstitially modified Sm_2Fe_{17} and $Sm_3(Fe,M)_{29}$ compounds:

(1) conventional casting followed by powder metallurgy [160-162]

(2) conventional casting followed by HDDR treatment [158,159]

(3) melt spinning followed by heat treatment,

(4) mechanical alloying followed by heat treatment [163],
Charging of the material is subsequently performed in N_2, NH_3 or hydrocarbon gasses at temperatures well below the decomposition temperature of the interstitial solutions into Sm nitride (or carbide).

(5) mechanical grinding of cast and homogenised Sm_2Fe_{17} in hydrocarbons or pyrazine [164].

The coercivities of $Sm_2Fe_{17}N_x$ powders have magnitudes more than sufficient for permanent magnet applications ($\mu_{oJ}H_c$ = 4.36 T, [163]) and also the remanences reach high values under carefully controlled processing conditions (B_r = 1.3 T; [165]). Similar promising hard magnetic properties were reported by Nasunjilegal et al. [146] for interstitial nitrides of the type $Sm_3(Fe,Ti)_{29}N_x$ and by Pan et al. [166] for $Sm_3(Fe,Mo)_{29}N_x$. For magnets prepared from ball-milled $Sm_3Fe_{24}Cr_5N_y$ a remanence of 0.87 T and an energy product of 105 kJ/ m^3 was reported Wang et al. [167].

Magnet bodies from nitrogenized Sm_2Fe_{17} powders can be prepared in the form of resin bonded magnets with BH_{max} = 136 kJ/ m^3, B_r = 9.0 T, $\mu_{oB}H_c$ = 6.5 T [168]. Magnet bodies made from nitrogenated compounds of more exotic compositions were investigated by Yamamoto et al. [169]. Arlot et al. [170] have milled $Sm_2(Fe,Co)_{17}N_{2.9}$ powders in hexane containing small amounts of the surface active agent Aerosol OT. After milling and washing with hexane the powders were treated with a $Zn(C_2H_5)_2$ solution in hexane. Zn coated powders were subsequently obtained by photodecomposition of $Zn(C_2H_5)_2$ in ultraviolet light. The demagnetisation curves of several Zn coated powders and the corresponding resin bonded magnets are reproduced in Fig. 14. The values of the energy product of the bonded magnets are BH_{max} = 152 kJ/ m^3. In order to explore the favourably low temperature dependence of the coercivity in the form of magnet bodies suitable for high-temperature applications Rodewald et al. [160] and Kuhrt et al. [163] have investigated Sn and Zn bonded magnets. In these cases, however, the remanences were fairly low (B_r < 0.7 T). More details as to preparative conditions and magnetic properties of these compounds can be found in the reviews of Li and Coey [13] and Fujii and Sun [18]. Gebel et al.[162] reported on the beneficial influence of addition of small amounts of Zr which makes the time-consuming homogenization treatment of Sm_2Fe_{17} alloys superfluous.

3.5. TERNARY IRON RICH COMPOUNDS OF THE TYPE $RFe_{12-x}M_x$

Besides the magnetic materials discussed in the preceding sections, Fe-based magnetic materials consisting mainly of $ThMn_{12}$-type compounds have occasionally attracted attention. A review of the $ThMn_{12}$-type materials has been given by Li and Coey [13], Buschow [171] and Fujii and Sun [18].

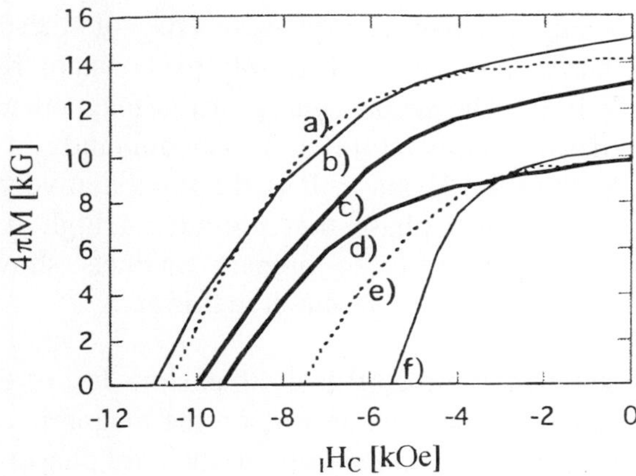

Fig. 14. *The demagnetisation curves of several Zn coated powders and the corresponding resin bonded magnets. Reproduced from Arlot et al. [170].*

The advantage of these materials is their good corrosion resistance, but it is difficult to attain sufficiently high coercivities for application in sintered permanent magnets. Permanent magnets based on $ThMn_{12}$ type compounds of the composition $Sm_{1-x}M_x(Fe,Co)12$, where M = Zr, Hf or Bi and $0.1 < x < 0.2$. were investigated by Ohashi [178]. Values of $\mu_0\,_jH_c$ up to 0.72 T were obtained on sintered magnets.

It is well known that the sign of the anisotropy in $ThMn_{12}$ type compounds can be changed by nitrogenation (see the reviews of Buschow [171], and Fujii and Sun [18], and references cited therein). This makes Nd based compounds suitable for hard magnetic materials. Mechanically alloyed and nitrogenated powders of the composition $Nd_{10}Fe_{75}V_{15}N_x$ seem to be the most favourable of this group with $T_c = 768$ K, M_s (295 K) = 131 Am^2/kg, and $\mu_{0j}H_c = 0.75$ T [172]. Zink-bonded permanent magnets of $NdFe_8Co_3N_x$ have been prepared by Inoue and Suzuki [173]. Particularly high values of the room temperature magnetization can be reached in $NdFe_{11.5}Mo_{0.5}N_x$ by substituting Y and Co for part of the Nd and Fe, respectively [174].

3.6. NANOCRYSTALLINE RARE EARTH BASED PERMANENT MAGNET

3.6.1. TWO-PHASE COMPOSITE MAGNETS

Two-phase materials can have a most interesting coercivity behaviour under special circumstances. Such behaviour has been described by Kneller and Hawig [175], who investigated the combined effect of two suitably dispersed and mutually exchange-coupled magnetic phases. One of these phases (k) is magnetically hard, has a large uniaxial anisotropy constant K, and is able to generate a high coercivity. The second phase (m) is magnetically soft. It has a larger magnetic ordering temperature and concomitantly a larger average exchange energy (A) than the k phase. It is the comparatively high saturation magnetisation (M_s) of the soft phase that provides a high remanence to the composite magnet. The possibility to prepare magnets showing remanence enhancement has led to an extensive research in this area.

Kneller and Hawig showed that the critical dimensions b_{cm} of the soft magnetic phase in exchange enhanced magnets depend on the magnetic coupling strength of the soft phase and the magnetic anisotropy of the hard phase:

$$b_{cm} = p \left(A_m / 2K_k \right)^{1/2} \tag{26}$$

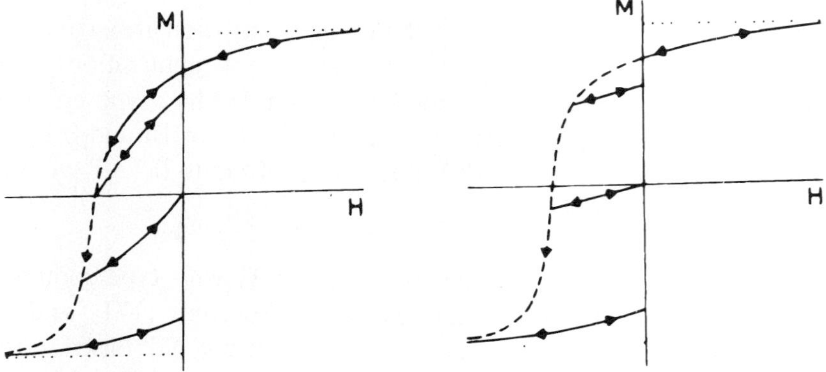

Fig. 15. *Demagnetizing curve for an exchange spring type magnet consisting of a hard magnetic and a soft magnetic phase (left), and a conventional magnet consisting of a single hard magnetic phase. From Kneller and Hawig [175].*

More details of the exchange enhancement mechanism can be found in the review of Buschow [20] and in the original paper of Kneller and Hawig [175]. Here we will restrict ourselves to describing several salient features of exchange enhanced magnets.

The hysteresis loop characteristics in an exchange enhanced composite magnet is illustrated in Fig.15 where it can be compared with the loop behaviour in a conventional quasi single phase magnet. The prominent feature of the former loop is that substantial portions of the demagnetizing curve can be traversed reversibly even though these curves are far from linear.

This offers substantial technological advantages offered by these so-called exchange spring magnets. This advantage can most easily be assessed by comparing the loop behaviour of the exchange spring magnet with that of a conventional magnet, shown in the right part of the figure. For high demagnetizing fields the irreversible magnetic losses can become considerable in non-ideal conventional permanent magnets, whereas they are fairly modest in the case of the exchange spring magnet. Kneller and Hawig draw attention to the fact that the magnetic behaviour of such materials is not only characterized by a reversible demagnetization curve ,implying maximum recoil permeability, but that one may also obtain an unusually high isotropic remanence-to-saturation ratio ratio, $m_r = M_r/M_s > 0.5$. Note that for an assembly of non-interacting randomly oriented crystallites with uniaxial anisotropy one would have expected a remanence-to-saturation ratio not higher than 0.5. Both these technologically important features were realised by Kneller and Hawig on melt spun and annealed $Nd_{3.8}Fe_{73.3}V_{3.9}B_{18}Si$ which had a remanence of 1.2 T and displayed exchange spring loop behaviour.

The technological realization of such composite permanent magnet materials has been proposed by Kneller and Hawig to be based on the following metallurgical principle: both the hard and the soft phases must be crystallographically coherent, a necessary condition for mutual exchange coupling. Therefore both phases must emerge from a common matrix phase. Most investigations reviewed in Ref. [20] did not deal with systems in which the requirement of crystallographic coherency was satisfied. Most of the systems for which remanence enhancement has been reported are permanent magnets in which the magnetically soft phase is α-Fe or an Fe-rich or Co-rich alloy. Examples of magnetically hard phases are $Nd_2Fe_{14}B$, $Sm_2Fe_{17}N_3$, Sm_2Co_{17} and $Nd(Fe,Mo)_{12}N_x$. The microstructures of all these composite magnets had in common that they consisted of a very fine distribution of the magnetic particles, falling into the nanometer range. In order to reach this fine distribution various techniques were employed, including melt spinning and mechanical alloying [20]. The effect of various kinds of substitutions T = Fe,V,Si,Ga,Cr on the magnetic properties of mechanically milled nanocrystalline $Nd_8Fe_{87}TB_4$ alloys was studied by Miao et al. who reported that it does not lead to significant improvements [176].

Computer simulations based on finite element analyses on composite systems composed of exchange coupled hard and soft phases have been made by Fukunaga et al. [177] and Schrefl et al. [72] Schrefl and Fidler [76,77]. Generally, these simulations have confirmed that the magnetic properties of isotropic nanostructured magnets are extremely sensitive to the microstructure. For instance, Schrefl et al. [72] reported that a mean grain size below 20 nm and a homogeneous microstructure with a very narrow grain size distribution are required for reaching a significant remanence enhancement and for preserving the high coercivity in $Nd_2Fe_{14}B$ based magnets. The calculations have also shown that the volume fraction of the magnetically soft phase can be as high as 50% without significant reduction of the coercivity for a grain size of about twice the domain wall width of the hard magnetic phase [77].

The presence of sharp corners and their consequences in micromagnetic calculations was discussed by Rave et al [179]. These authors showed that no effect on rounded corners on coercivity is to be expected as long as the rounding radius is small compared to the exchange length.

Zern et al. [180], Willcox et al. [182] and Miao et al. [183] have investigated the relationship between coercivity and remanence and the relative fraction of nanocrystalline α-Fe in nanocrystalline NdFeB type magnets. Optimally heat treated samples were reported by the latter authors to display a remanence-to saturation ratio that increases with increasing volume fraction of α-Fe, while the coercivity decreases. This is in agreement with results of Schrefl et al. [73,184] who performed numerical micromagnetic calculations on exchange coupled nanocrystalline materials and found that with increasing volume fraction of the magnetically soft phase the remanence and coercivity should increase and decrease, respectively.

3.6.2. QUASI SINGLE-PHASE NANOCRYSTALLINE MAGNETS

It was mentioned already in section 3.1.2 that, in contradistinction to $Nd_2Fe_{14}B$ powder obtained by milling or grinding of normally cast alloys, powder obtained from melt spun alloys is coercive and can therefore be used for bonded magnets. Under optimal condition, the melt spinning process leads to microstructures in which the grains are oriented at random and in which the mean crystallite size is around 60nm [185], or even much lower [180]. Although the small grain size leads to a sufficiently high coercivity, the random grain orientation does generally not lead to remanences higher than $m_r = M_r/M_s = 0.5$. However, higher remanence ratios can be attained when the processing conditions are modified in a way that smaller mean crystallite sizes are reached. This effect has been attributed by Manaf et al. [186] and Davies [187] to exchange coupling between the moments in neighbouring crystallites. A TEM investigation in which exchange coupled and decoupled NdFeB and PrFeB magnets are compared has been presented by Zern et al. [180] and Goll et al. [181], respectively. Analysis of these materials in terms of the nucleation model (eq. 25) has revealed that there are quite marked differences in microstructural parameters between exchange coupled magnets and conventional magnets.

It is interesting to note that a similar opposite behaviour between remanence and coercivity is found in these quasi single-phase nanocrystalline alloys (containing basically only the hard magnetic phase, or containing comparatively only small amounts of the second phase) as in the two-phase composite alloys discussed in the preceding section. For the quasi single-phase nanocrystalline alloys Huo and Davies [188] used the random anisotropy model in conjunction with an average intergrain interaction and showed analytically that the remanence will be increased at the expense of the coercivity when the exchange interaction between the grains increases. Also the numerical micromagnetic calculations of Schrefl et al. [73] for quasi single-phase nanocrystalline alloys confirmed such a relationship, showing increasing remanence enhancement and decreasing coercivity as a function of decreasing average grain diameter. Davies et al. [187,189] investigated also Ga containing quasi single phase NdFeB alloys. In these materials the influence of decreasing grain size on the remanence and coercivity is different since both quantities were found to increase with decreasing grain size, leading to an excellent combination of properties.

4. 3d-TRANSITION METAL ALLOY MAGNETS

4.1. ALNICO-TYPE ALLOYS

The alnico alloys are composed of Fe, Co, Ni and Al and various types of additives. The hard magnetic properties of the alnico alloys can vary quite substantially, depending on their composition and heat treatment. It has become customary to classify the alnico alloys by numbers, ranging from 1 to 9.

The alnicos 1-4 are isotropic magnets of relatively low Co content (0-20 wt%). Much higher Co concentrations (24-40 wt%) are used in the alnicos 5-9. These magnets also contain considerable concentrations of Ti (5-8 wt.%) and are manufactured in a way that makes them anisotropic.

The first step in the manufacturing of alnico magnets is melting, which is mostly carried out in high frequency furnaces. A large advantages of alnico alloys is that the melting can be performed in air, although sometimes an apron of inert gas is blown over the melt. The alnico alloys can also be prepared by sintering powders of the raw materials, Al being always added in the form of a prealloy with one ore more of the other constituent metals.

In contrast to the rare earth based permanent magnet materials discussed in the preceding sections the alnico alloys have a rather insignificant magnetocrystalline anisotropy. They owe their magnetic hardness to shape anisotropy which implies that the microstructure of the alloys is of prime importance. In fact, the alnico magnets can be characterized as fine-grained alloys in which elongated ferromagnetic particles are dispersed in a basically nonmagnetic matrix. This particular microstructure is reached in a heat treatment during which the origially homogeneous alloy separates into two phases via a so-called spinodal decomposition.

It has been mentioned already that all alnico alloys are multicomponent systems. All these systems have in common that they form homogeneous bcc solid solutions at high temperatures (α phase). At lower temperatures there is a miscibility gap. This means that upon cooling the homogeneous α phase will separate into two phases of different composition, α_1 and α_2, where α_1 represents a strongly ferromagnetic bcc phase rich in Co and Fe and α_2 a nonmagnetic or only weakly ferromagnetic bcc phase rich in Al and Ni.

After melting, the alnico alloys are subjected to a homogenization treatment at about 1250°C at which temperature the alloys still consists of a single phase (α). Subsequently the alloys are annealed at lower temperatures so that spinodal decomposition can occur. The spinodal decomposition of the α phase into the α_1 and α_2 phases can take place spontaneously, but is diffusion-limited. It can proceed at a sufficient rate only at relatively high temperatures (800-850°C). Initially, the concentrations of the Fe(Co) atoms show a periodic variation (assumed to be sinusoidal after decomposition has set in) and the amplitude of the composition fluctuations increases with time until the phase separation into α_1 and α_2 is complete. This heating step is generally terminated after the formation and subsequent growth of the particles to their final shape and size has taken place. Particle growth is anisotropic, because the interfacial energy depends on the crystallographic orientation of the boundaries between the α_1 and α_2 type particles.The resulting microstructure consists therefore of a fairly complicated three-dimensional interlocking system of elongated α_1 particles having their axis of elongation approximately parallel to the crystallographic <100> directions.

A substantial improvement in magnetic properties can be reached during a further tempering treatment, mostly performed at about 600°C for several hours. During this low-temperature aneal a continuous change in the composition takes place caused by the diffusion of Fe and Co atoms to the ferromagnetic particles, which increases the difference in saturation magnetic polarization between the Fe(Co)-rich particles and the surrounding matrix (Ni-Al-rich). It is important to realise that the spinodal decomposition alone does not produce a sufficiently large shape anisotropy in the ferromagnetic α_1 phase particles. This is because the difference in the saturation magnetizations between the initial α_1 particles and the matrix phase α_2 is relatively small so that the effective shape anisotropy field of the particles, being proportional to $J_s(\alpha_1) - J_s(\alpha_2)$, is also small in spite of the elongation. The heat treatment at 600°C is therefore desirable in order to increase the magnetization anisotropy, which makes it possible to obtain the high coercivities and the optimum permanent magnet properties.

It has been mentioned already that particle growth is anisotropic, which results in an elongation parallel to the <100> directions. Substantial improvements in magnetic properties can generally be obtained by controlled cooling of the alloys after the homogenization treatment at about 1200°C to about 800°C in a saturating magnetic field, as in anisotropic alnico 5 alloys. This thermomagnetic treatment promotes the formation of magnets in which the easy magnetization direction of the grains formed during the spinodal decomposition is parallel to the direction of the magnetic field applied during cooling. The elongation of the particles in the field direction originates from the lowering of the magnetic free energy of the particles when the axis with the lowest demagnetizing factor is along the direction of the applied field.

It can be envisaged that even better magnetic properties can be obtained with the thermomagnetic field applied parallel to one of the <100> directions of a properly oriented single crystal. In practice, so-called columnar crystallized alnico alloys are applied instead of expensive single crystals. Such alloys can be obtained by a grain orienting process before the thermomagnetic and tempering treatment. A common procedure is to cast the alloys in heated or exothermic molds onto water-cooled steel or copper slabs. During solidification of the alloy on the cold surface the grains tend to grow with their long axis parallel to the <100> directions, preferentiaaly perpendicular to the cold surface. This leads to a semicolumnar alloy in which the columnar axis is parallel to one of the <100> directions, for instance, parallel to [001]. Alloys manufactured in this way are often referred to as alnico DG (directed grain). The thermomagnetic treatment for this type of alloy, for instance for alnico 5DG, is applied with the magnetic field parallel to the [001] direction. As mentioned already above, the magnetic properties in the easy magnetization direction can be further improved by subsequent tempering for several hours at about 600°C.

The relatively high coercivities and remanences in the alnico alloys originate from the shape anisotropy associated with the elongated Fe(Co)-rich particles imbedded in a nonferromagnetic matrix.. When assuming that the ferromagnetic particles are approximately prolate ellipsoids of revolution, the Stoner-Wohlfarth theory predicts that the coercivity is proportional to the saturation polarization J_s of the Fe(Co)-rich particles and also proportional to a factor related to the difference in the effective demagnetization factors perpendicular ($N_\perp$) and parallel ($N_{||}$) to the preferred direction of magnetization in the particles:

$$H_A = H_c = 1/\mu_o\, f(q)[\, N_\perp - N_{||}\,]\, J_s \qquad (27)$$

where $f(q)$ is an averaging factor taking account of the non-ideal orientation of the preferred axes of the particles with respect to the direction in which H_c is measured. For the case that the particles are non-interacting uniaxial single-domain particles arranged at random one has $f(q) = 0.5$.

For highly elongated particles the factor $f(q)$ may approach the value 1. In the case of spheroid particles there is a considerable difference in demagnetizing factor for particles magnetized perpendicular and parallel to the flat surface of the spheroid. In the limit of extremely flat and elongated spheroid particles one has $[\, N_\perp - N_{||}\,] = 1 - 0 = 1$. In such materials the coercivity would reach an upper limit equal to $H_c = J_s/\mu_o = M_s$. With $M_s = 1.7$ MA/m this upper limit becomes $H_c = 1.7$ MA/m. Even for magnets in which $[\, N_\perp - N_{||}\,] = 0.5$ the coercivity would become 850 kA/m. The latter value is much higher than the actual ones found in alnico materials, which is primarily attributed to the less perfect shape of the thin ferromagnetic particles and also to the fact that the matrix is magnetic to some extent.

One of the favourable magnetic properties of the alnico alloys is the very high Curie temperatures (700 to 850°C). The small negative reversible temperature coefficient, $\alpha(B_r) \approx -0.02\%$ per °C leads to an excellent flux stability at elevated temperatures. The alnico alloys are chemically and metallurgically very stable. In fact, alnico 5 is the only magnet material that has some long-term utility at temperatures up to 500°C.

As may be derived from the demagnetizing curves shown for two representatives of alnico magnets in Fig. 16, a drawback of the alnico alloys is their low coercivity in comparison to the rare earth based magnets described in the previous sections. The non-linear behaviour of the B(H) curve in the second quandrant is a serious disadvantage in device design and dynamic operation. It limits the attainable energy product in spite of high remanence, which is comparable to that of NdFeB type magnets.

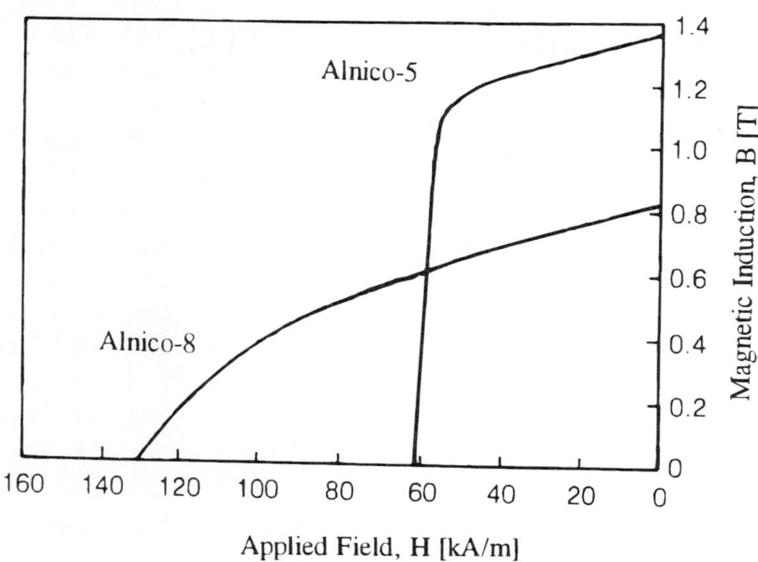

Fig. 16. *Examples of the demagnetizing behaviour of two typical alnico alloys*

Many decades ago the so-called alnico alloys formed the basis for the majority of permanent magnets manufactured worldwide. At present production of alnico type magnets is fairly low but still substantial, owing to the excellent temperature characteristics mentioned. The properties of alnico magnets are best suited to magnetic circuits of high permeance because the maximum energy product corresponds to permeance ratios larger than five. In practice, this means that alnico magnets have to be comparatively long in the easy direction and their cross sectional area should be relatively small. Application areas include sensors, electron tubes, Watt hour meters, and minor applications in the field of communication and separation. Their application in motors and generators is restricted to those operating at high temperatures. More details of alnico magnets can be found in reviews written by McCurrie [190] and McCaig and Clegg [191].

4.2. Fe-Cr-Co ALLOYS

The hard magnetic properties of Fe-Cr-Co alloys have roughly the same origin as that of alnico alloys. It is due to shape anisotropy associated with an extremely fine-scale two-phase microstructure. This microstructure can be generated by profiting from the presence of a miscibility gap. Heat treatments performed at temperatures within the miscibility gap lead to spinodal decomposition of an Fe-rich high-temperature α-phase (bcc) into a strongly ferromagnetic α_1 phase and an almost nonmagnetic Cr-rich α_2-phase. The spinodal decomposition temperature and the concomitant decomposition rate in pure Fe-Cr alloys is much too low for practical purposes and one of the reasons for using Co additions is to increase the decomposition temperature.

Permanent magnets based on Fe-Cr-Co alloys were discovered in 1971 by Kaneko et al. [192]. The initial alloys were still comparatively rich in Co. The composition suggested by Kaneko et al. was 30% Cr, 25% Co and balance Fe, with optional additions of Mn or Si.. The heat treatment recommended consisted of a solution treatment between 1300°C and 1350°C in argon, followed by a rapid quench, 30 min. at 630 to 640°C in a magnetic field, and a final heat treatment without a field for up to 6 hrs at 600 and 560°C. The best properties claimed were $B_r = 1.3$ T, $H_c = 46$ kA/m and $(BH)_{max} = 42$ kJ m^{-3}.

The Fe-Cr-Co alloys have permanent magnetic properties comparable to those of alnico alloys. They have, however, the large advantage of being ductile and containing less of the comparatively expensive Co. The standard route for manufacturing Fe-Cr-Co magnets involves induction melting or arc melting. Subsequently, the cast ingots are hot rolled and cold rolled into the desired shape. All these treatments are preferably performed at temperatures above the miscibility gap mentioned above. They are followed by a solution annealing treatment in the range 900-1000°C, i.e. still within the single a phase region. It is important to retain the homogeneous nature of the material by quenching to room temperature. The favourable microstructure and the associated magnetic hardness is obtained from the single-phase quenched material by giving it a heat treatment at temperatures located within the miscibility gap. Often a two-stage cooling method is applied. The initial fast cooling leads to the optimal particle size, whereas the second more slowly cooling leads to an optimum compositional difference between the two coexisting phases.

The magnets obtained by the single stage or two-stage procedure described above are isotropic, with microstructures characterized by fine slightly elongated particles of the α_1 phase embedded in a relatively Cr-rich almost non-magnetic α_2 phase. Two techniques are commonly employed to obtain anisotropic magnets. The first technique involves deformation-ageing. It takes advantage of the comparatively large ductility of the Fe-Cr-Co alloys still present after partial spinodal decomposition. After cooling through the spinodal decomposition temperature at a rate favouring oversized and spherical particles the alloy is subjected to anisotropic mechanical deformation. The latter may consist of extrusion or rolling which elongates and aligns the particles. After a sufficiently high degree of particle elongation has been reached, the alloy is given a further heat treatment aiming at optimizing the compositional difference between the magnetic particles and the nearly non-magnetic matrix, as was discussed already in the section dealing with alnico alloys. Anisotropic Fe-Cr-Co magnets obtained by deformation ageing can have energy products up to $(BH)_{max} = 78$ kJ m^{-3}, with $B_r = 1.30$ T and $H_c = 86$ kA/m.

The second technique uses magnetic field ageing. In the same way as described already for the alnico alloys in section 4.1, it is possible to obtain an anisotropic microstructure in Fe-Cr-Co alloys by giving them a heat treatment in the presence of a magnetic field (typically 0.5 - 2 kA/m). This field leads to an anisotropic growth of the α_1 particles. As for the alnicos, such treatments may include materials having a cast columnar structure favouring growth in the <001> direction. Generally this thermomagnetic heat treatment enhances the remanence, the shape anisotropy and the concomitant coercivity. Anisotropic Fe-Cr-Co magnets obtained by magnetic field ageing of directionally solidified alloys can have energy products up to $(BH)_{max} = 76$ kJ m^{-3}, with $B_r = 1.54$ T and $H_c = 67$ kA/m.

The Fe-Cr-Co alloys are less brittle than alnico alloys. Therefore the former do not require casting or grinding into final shape. Generally they can be applied in all cases where a higher ductility is required than in alnico type alloys, because they are amenable to comparatively easy cold formability into complicated shapes. The most prominent large scale applications of Fe-Cr-Co alloys is in telephone receivers as is described by Jin et al. [193,194] For more details of the magnetism of Fe-Cr-Co alloys and their manufacturing the reader is referred to reviews written by McCaig and Clegg [191], Jin and Chin [195], and Homma [196].

5. MISCELLANEOUS MAGNET MATERIALS

5.1. PERMANENT MAGNETS BASED ON NOBLE METAL COMPOUNDS

The Pt-Co phase diagram is characterised by a range of complete solid solution in the solid state at high temperatures. The corresponding crystal structure is a disordered fcc structure, in which the crystallographic sites are statistically occupied by Co and Pt atoms. This high-temperature phase is of little use for permanent magnet purposes because it does not show a sufficiently high magnetic anisotropy. However, structural transformations take place when the rapidly cooled $Co_{1-x}Pt_x$ alloys are annealed at lower temperatures. For high Pt concentrations ($x > 0.75$), the fcc phase gives rise to a disorder-order transformation. For relatively low Pt concentrations ($x < 0.23$), the disordered fcc phase transforms into an hcp structure. The applicability of $Co_{1-x}Pt_x$ alloys as permanent magnets is based on the transformation of the disordered fcc phase to an ordered face-centred tetragonal phase (fct) in the intermediate concentration range ($40 < x < 75$), because only the latter phase has a sufficiently high uniaxial magnetic anisotropy, the easy magnetization direction being along the tetragonal c-axis. The origin of the magnetic anisotropy and the role played by the orbital moment in transition metal compounds in CoPt was studied by Daalderop et al. [197] by means of first principles band structure calculation using the local spin density approximation. For a more detailed discussion of this topic, the reader is referred to the review of Buschow [17] or to the original article of Daalderop et al. The Curie temperature of these materials is around 500°C for the equiatomic composition and slightly decreases with Pt concentration.

The maximum of the magnetic anisotropy energy $E_A = K_1 + 2 K_2$ is located at the equiatomic composition, but the maximum of $H_A = (2 K_1 + 4 K_2)/ M_s$ is located at slightly higher x values because of the strong decrease of the saturation magnetization M_s with x. For obtaining optimum values for the coercivity slightly higher Pt concentrations than the equiatomic composition would be desirable because it is the anisotropy field rather than the anisotropy energy that determines the coercivity. However, the energy product depends very strongly on the magnetisation. For this reason concentrations close to the equiatomic composition are generally chosen for practical applications as permanent magnets.

Co-Pt alloys are known to have outstanding mechanical strengths and for this reason the powder metallurgical route described in section 3 (see Fig. 8) is not well applicable. In fact, the generation of coercivity by particle comminution is not required in this case because the presence of the fcc-to-fct phase transition provides ample means of obtaining sufficiently small fct particles. In other words, the generation of coercivity in CoPt alloys is a matter of controlling the nucleation and growth of fct particles during the phase transformation. This process involves heat treating the ingots under carefully selected conditions.

Investigation made by Kaneko et al. [198] have shown that the coercivity passes through a maximum when CoPt alloys are heat treated for variable ageing times at temperatures sufficiently below the fcc-fct transformation temperature, with an optimum for an annealing temperature around 680 °C. Kaneko et al. furthermore found that even better results are obtained when the first ageing step is followed by a second ageing step, provided the first ageing step is kept sufficiently short to prevent overageing. The first annealing step was assumed by Kaneko et al. as being primarily responsible for the formation of a fine precipitate of the fct phase whereas the second step was held responsible for an increase of the magnetocrystalline anisotropy of the fct phase. The latter increase is probably due to more complete atomic ordering of the Co and Pt atoms in the fct grains. By applying the two-step annealing process, Kaneko et al. obtained coercivities of more than 700kA/m on alloys annealed first at 680°C for various times and subsequently at 600 °C for variable times. The corresponding energy products of the ingot magnets can reach values around 100 kJ/ m^3.

A behaviour fairly similar to that described above for $Co_{1-x}Pt_x$ alloys can be found also for $Fe_{1-x}Pt_x$ alloys. High coercivities for $Fe_{1-x}Pt_x$ alloys were obtained when alloys around the equiatomic composition were first homogenized at high temperatures and then annealed at temperatures below the order-disorder transition temperature. Small amounts of Nb lead to magnets of hard magnetic properties superior to those of the binary $Fe_{1-x}Pt_x$ alloys [199].The temperature dependences of the residual flux densities of several Fe-Pt based alloys are shown in Fig. 17.

Detailed investigations of the microstructure of heat treated $Fe_{1-x}Pt_x$ alloys and the concomitant coercivities were made by Tanaka et al. [200]. These authors showed that composition control is essential for the attainment of high coercivities. In as-quenched alloys optimum coercivities were reached for 39.5 at.% Pt because the disordered fcc phase disappears during quenching and can no longer act as nucleation centers for domain walls.

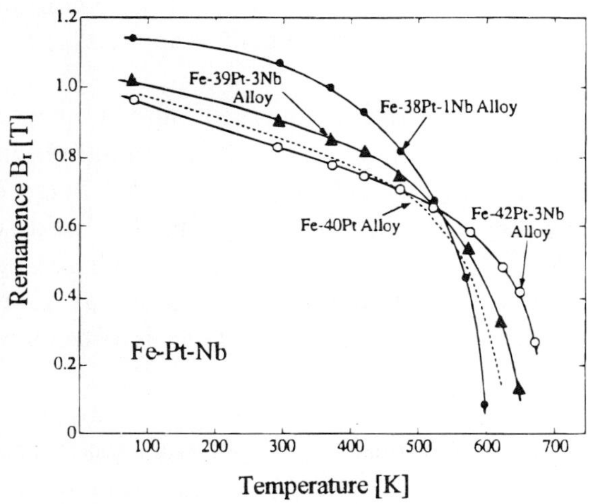

Fig. 17. *Temperature dependences of the residual flux densities of several Fe-Pt based alloys. From Watanabe [199].*

Tanaka et al. [200] furthermore showed that alloys with optimum coercivities are reached after heat treatments in which the local stress associated with the cubic-to-tsetragonal transformation is not yet released by twin formation, and the fct phase is still present in nanoscale antiphase domains. The antiphase domain boundaries between the single domain particles act as pinning sites for domain wall motion and promote high coercivities. Too long ageing treatments lead to particle growth so that these are no longer single domain particles and the coercivity becomes reduced accordingly.

Investigations made by Watanabe [199, 201] have shown that magnetic properties even superior to those of Co-Pt alloys can be obtained on Fe-Pt alloys containing small amounts of niobium. The ingot was first homogenized at 1325°C and quenched in water. A subsequent isothermal annealing process performed in a temperature range between 600 and 700°C leads to the desired high coercivity. The Fe moments are somewhat higher than the Co moments in these Pt alloys. For this reason the remanences in the Fe alloys are higher than in the Co alloys, favouring an equally larger maximum energy product with $(BH)_{max}$ values of more than 160 kJ/ m^3

The Pt based magnets are extremely expensive owing to the fact that they consist of roughly 75 weight% platinum. Advantages of these magnets are that they can be produced as ingot magnets, avoiding the complicated powder metallurgical manufacturing route. The magnets are of high mechanical strength and of unequalled corrosion resistance. Because of price considerations they are produced only in small quantities, and are mainly applied for medical implants.

When comparing the manufacturing routes of alnico magnets and Pt alloy magnets one may notice that there is a striking similarity. In both cases the manufacturing benefits from the fact that an extended range of solid solution exists at high temperatures and that upon cooling this solid solution becomes supersaturated and leads to the precipitation of new phases. In both cases the decomposition of the supersaturated solid solution has to be performed at sufficiently low temperatures so that the resulting microstructures consist of fine grains and exhibit magnetic hardness.

Interestingly, neither of the two parent solid solutions has a magnetic anisotropy of any significance. In the case of the Pt alloys the required magnetic anisotropy is obtained by the formation at low temperatures of a phase of lower symmetry (fct) than the parent phase (fcc). In contrast to the latter, the former phase exhibits a fairly strong magnetocrystalline anisotropy. Both phases have the same composition and the particle size is not very critical for the generation of anisotropy. Roughly speaking it can be said therefore that the main goal of finding optimum annealing treatments is to generate microstructures with grain sizes sufficiently small for generating coercivity. By contrast, the main goal of finding optimum annealing treatments in the case of alnico alloys is to produce microstructures where not only the size but also the shape of the precipitated particles is at a premium, because not only coercivity but also shape anisotropy has to be generated due to the absence of magnetocrystalline anisotropy.

5.2. MAGNETS BASED ON τ- Mn-Al

Permanent magnets obtained from Mn-Al alloys all contain the so-called τ-phase. This is a compound with an f.c.tetragonal structure (CuAu-type superstructure), occurring in the composition range 51-58 at% Mn, or equivalently 67-73 wt.% Mn. The Mn atoms occupy the (0,0,0) sites in this crystal structure and their moments couple ferromagnetically. Because the composition is richer in Mn than corresponding to the equiatomic composition not all the Mn atoms can be accommodated at the (0,0,0) sites. This implies that the excess Mn atoms have to share the (1/2,1/2,1/2) sites with the Al atoms. Braun and Goedkoop [202] showed that the moments of the excess Mn atoms couple antiferromagnetically with the of the Mn moments in the main Mn sublattice. As will be discussed below, this has important consequences not only for the saturation magnetization but also for the hard magnetic properties of permanent magnets based on Mn-Al alloys.

A prominent feature of the τ-phase is its metastable nature. This means that this phase is not found in the Mn-Al phase diagram (see for instance, the phase diagram of Liu et al. [203], shown in Fig. 18). It can be prepared from the high-temperature equilibrium phase e occurring in this concentration range, that has a hexagonal close packed structure. The formation of the fct τ-phase from the hcp ε-phase is not straightforward. Investigations of Vlasova et al. [204] and Van den Broek et al. [205] have shown that the formation of the τ-phase from the ε-phase proceeds via an intermediate phase of orthorhombic structure (B19), or via an intermediate phase of fcc structure. This leads to one of the following two reaction schemes:

$$\text{h.c.p.}(\varepsilon) \Rightarrow \text{B19}(\varepsilon') \Rightarrow \text{f.c.t.}(\tau)$$

$$\text{h.c.p.}(\varepsilon) \Rightarrow \text{f.c.t} \Rightarrow \text{f.c.t.}(\tau)$$

where the τ phase is formed as the end product of successive transformations. Most likely the high-temperature hexagonal-closed-packed structure (ε) transforms first into the orthorhombic ε'-phase by an ordering reaction. The metastable face-centered-tetragonal ferromagnetic τ-phase is subsequently formed by way of shear transformation, or by a massive trasformation as advocated by Hoydick et al. [206]. The annealing time has to be chosen carefully because the metastable τ-phase, after a time that depends on alloy composition and the annealing temperature, tends to decompose again into the two stable phases with β-Mn and Cr_5Al_8 type structure. Heat treatments proposed to obtain and preserve the τ-phase have been discussed in more detail in Ref. [20].

Several investigations have shown that substantial improvements with respect to the formation and decomposition kinetics of the τ-phase can be reached by doping with carbon (Kojima et al. [207], Ohtani et al. [208], Pareti et al. [209]). Although the C doping has a beneficial effect on the saturation magnetization, there is a substantial decrease in Curie temperature.

The partial occupation of Mn atoms of Al sites at (1/2,1/2,1/2) implies that many antiphase domain boundaries occur in the crystal lattice of the τ-phase. As has been discussed in detail elsewhere [20], these antiphase domain boundaries can act as nucleation sites for Bloch walls, which leads to relatively easy magnetization reversal and hence to low values of the coercivity and remanence.

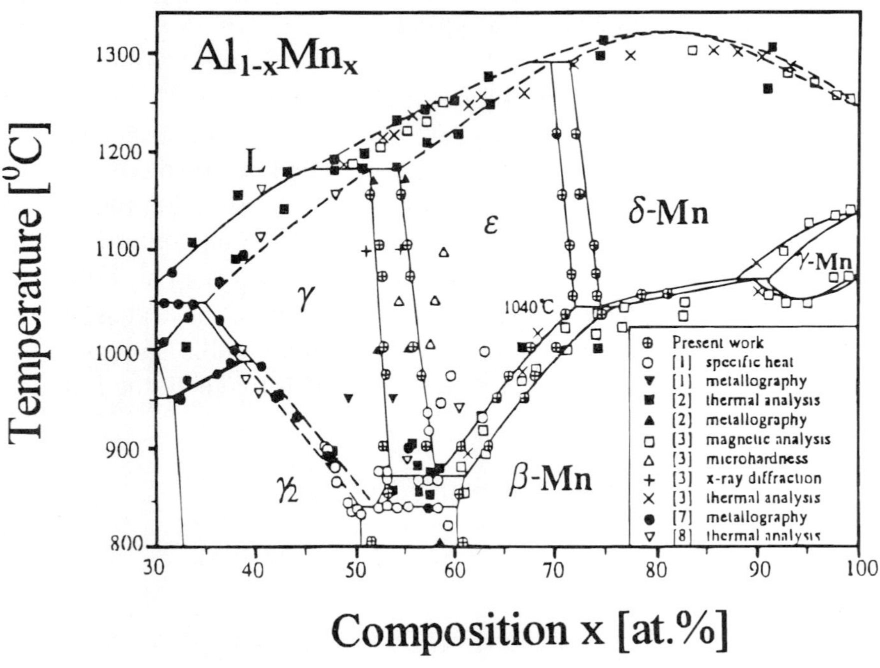

Fig. 18. *Mn rich portion of the Mn-Al phase diagram. This phase diagram was constructed from data obtained by Liu et al. [203], marked as present work in the figure. The remainder of data were obtained with various techniques by other authors cited in the report of Liu et al.*

Ohtani et al. [208] showed that the hard magnetic properties can be significantly improved if the magnets are made by a high-temperature extrusion process. This process leads to microstructures that are characterized by very fine grains and in which the number of antiphase boundaries is strongly reduced. First the alloy (typically 70.0 wt% Mn, 29.5 wt% Al) is homogenized at 1100°C for 1 hour. After quenching to 500 °C the alloy is annealed at 600°C for 30 min. Hot extrusion is performed at 700°C with a pressure of 80 kg/mm^3. Finally the extruded material is aged at 700°C for 10 min. The energy products of the hot-extruded magnet can reach values close to 56 kJ/ m^3 (7 MGOe). Although this is a comparatively low value one has to take into consideration that the raw materials cost are low and the same holds with respect to the production cost, the whole processing route can be performed in air. Further improvements in the manufacturing process, proposed by Yamaguchi [211], involve the extrusion of gas atomized powders . The magnets are reported to have excellent mechanical strength, do not chip, are machinable and take on good surface finish.

6. HARD FERRITES

The ferrites form a large class of ceramic materials. In terms of technological applications one may distinguish between two main types of ferrites, soft ferrites and hard ferrites. The former materials have a cubic crystal structure and are mainly used in radio frequency transformers, magnetic heads and other electronic applications. These materials fall outside the scope of the present report. Progress made in the field of spinel ferrite research has been reviewed by Brabers [212].

The hard ferrites used for permanent magnet purposes have hexagonal crystal structures and for this reason are frequently called hexaferrites. The formula composition of these ferrites can generally be represented as $n(MeO)\times m(Fe_2O_3)$ where n and m are natural numbers and where Me is a divalent metal.

When inserting n=1 and m=6 in this formula one obtains the composition $MeO\times 6(Fe_2O_3)$ or $MeFe_{12}O_{19}$ representative for the so-called M type ferrites. The most common M type ferrites are those where Me = Ba, Sr or Pb.

For n=3 and m=8 one obtains the formula composition of the so-called W-type ferrites, $3(MeO)\times 8(Fe_2O_3)$. There are usually two (or more) types of divalent metals Me involved. The formula composition of the W-type ferrites is therefore most conveniently represented by $Me_1\underline{Me}_2Fe_{16}O_{27}$ .Well known examples of such compounds are formed with Me = Ba or Sr and $\underline{Me}$ consisting of Fe(II) and large amounts of Mg, Zn, Cu, Ni, Co or Mn.

Both types of hexaferrites owe their hard magnetic properties to their comparatively large magnetocrystalline anisotropy. These hard ferrites play a dominant role in the permanent magnet market which is mainly due to the low price per unit of available energy, the wide availability of the raw materials and the high chemical stability.

M-type ferrites can be regarded as the more common type of the hard ferrites. They adopt the magnetoplumbite structure characterized by close packing of oxygen and Me ions with Fe atoms at the interstitial positions. Alternatively one may describe this crystal structure as being built up of cubic blocks with the spinel structure and hexagonal blocks containing the Me ions. There are two formula units $MeFe_{12}O_{19}$ per unit cell and the Fe^{3+} ions are located at five nonequivalent crystallographic positions. The crystal structure of the W-type ferrites is somewhat more complex.

In the past many research efforts have been spent in modifying the magnetic properties of the hard ferrites by substitutions. It is important to bear in mind that substitutions are only possible when the rule of charge conservation is obeyed. A few examples of substitutionally modified hard ferrites satisfying the requirement of charge conservation are :

$$Ba_{1-x}Na_x^{1+}La_x^{3+}Fe_{12}O_{19} \quad \text{(M-type)} \qquad BaFe^{2+}Fe_{16}O_{27} \qquad \text{(W-type)}$$

$$BaFe_{12-2x}Ti_x^{4+}Co_x^{2+}O_{19} \quad \text{(M-type)} \qquad SrZn(LiFe)_{0.5}Fe_{16}O_{27} \quad \text{(W-type)}$$

In all these cases Fe has to be regarded as trivalent when not indicated differently. A detailed survey of various types of substitution in M-type ferrites is given in the review of Kojima [213] and Kools [214].

The Curie temperatures of the hexagonal ferrites are fairly high and fall into the temperature range between 700 and 750 K. Also the Fe moments in these ceramic materials adopt much higher values than the Fe moments in the Fe based intermetallic compounds discussed in the previous sections. For instance, each of the Fe^{3+} ions in $BaFe_{12}O_{19}$ carries a magnetic moment of 5 μ_B, which is more than twice the value found in the intermetallic compounds. In spite of this, the saturation magnetization of the ferrites is substantially lower than that of the intermetallic compounds. The reason for this is the fairly complicated magnetic interaction between the Fe moments. The moment of the Fe ions residing on the same crystallographic position are ferromagnetically coupled. However, the coupling between iron moments belonging to different crystallographic positions is ferromagnetic for some sites and antiferromagnetic for other sites.

All these magnetic interactions are determined by a superexchange mechanism, mediated by the oxygen atoms. As is illustrated in the survey of Kojima [213], there is a strong preference for ferromagnetic coupling when the angle Fe-O-Fe approaches 180 deg and the distance Fe-O-Fe is comparatively small. The resultant spin structure is ferrimagnetic, and the partly antiparallel coupling of the Fe moments leads to a net moment per unit cell of only 40 μ_B at 4.2 K. This considerably smaller than the value 12x5 μ_B = 60 μ_B that would have been obtained for ferromagnetic alignment. A more detailed description of the superexchange interaction is given in the review of Guillot [215].

Of particular interest in the hexagonal ferrites is the temperature dependence of the saturation polarization J_s that increases with decreasing temperature much more slowly than would be expected on the basis of the Brillouin function. This has as a consequence that the room temperature value of J_s in these materials is relatively low, much lower than the value corresponding to the saturation moment of 40 μ_B per formula unit mentioned above. It also implies that the temperature coefficient of J_s is fairly high (-0.2% K^{-1}).

The magnetocrystalline anisotropy in the hard ferrites is characterized by a comparatively high positive value of the first order anisotropy constant K_1. The higher order anisotropy constants (K_2, K_3, ...) are negligibly small. This situation corresponds to an easy magnetization direction along the c-axis. Because J_s decreases more strongly with temperature than K_1 in the lower temperature range one has here the rather unusual situation that the anisotropy field $H_A = 2$ K_1/ J_s first (slightly) increases with temperature before it eventually decreases. This, in turn, has important consequences for the coercivity of these materials in the temperature range of technological interest, because also H_c increases with temperature in this range (see section 2.3).

Sintered permanent magnets based on hard ferrites can be manufactured either by dry or wet pressing followed by sintering. A considerable amount of ferrites is also used in bonded magnets made by mixing ferrite powders with either rubber or plastic binders followed by compacting or molding. Of both types of magnets anisotropic as well as isotropic forms are commercially available. The former types are characterized by higher remanences owing to the magnetically induced alignment of the powder particles during processing. In the latter magnets the orientation of the powder particles has a random distribution.

The first step in the manufacturing process involves compound formation. Compounds of the M-type ferrites are usually made by calcination of appropriate mixtures of Fe_2O_3, $BaCO_3$ or $SrCO_3$ and minor additives at high temperatures (about 1250°C). After this so-called prefiring proces the resulting product is milled to very fine particles (below 1 mm). For obtaining bonded magnets the powder is mixed with resin, granulated and extruded (see also section 7). This manufacturing process is schematically represented in Fig.19. In more sophisticated manufacturing routes the milling is performed in water. The slurry can be dried by heating the powder in ovens or by spray drying. After mixing the dried ferrite powders with lubricants and small amounts of additives, the material is compacted by pressing (dry pressing). In order to obtain higher remanence values the pressing is frequently done in a magnetic field. However, because of the plate-like character or the particles the latter tend to stick together and this makes alignment in a field during dry pressing somewhat ineffective. The pressed magnet bodies are subsequently subjected to a firing treatment consisting of sintering in air at high temperatures (at about 1250°C).

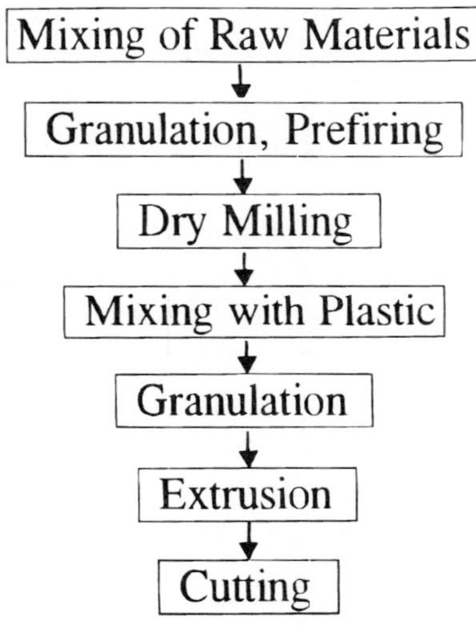

Fig. 19. *Schematic view of the manufacturing route of bonded ferrite magnets*

Wet pressing is used to manufacture anisotropic sintered magnets. For this process materials are used made by prefiring in the same way as described already above. However, the additives used are different from those used for dry pressing (SiO_2 and/or B_2O_3) in order to better control grain growth and densification. After prefiring the material is wet-milled with steel balls in water to obtain a thick suspension (slurry) in which the fine powder particles (preferably single crystals) have sufficient mobility to align themselves along the preferred magnetization direction in an external field during wet pressing. Teng et al. [216] used a combination of dry and wet milling. First, the powders are dry-milled to average particle sizes below 1.8 mm and subsequently wet-milled to

0.75 mm, using H_3BO_3 as dispersant to prevent particle cohesion and to promote alignment.

When the dry or wet pressed compacts are sintered in air anisotropic shrinkage occurs. This shrinkage is about 25% in the easy magnetization direction and about 15% in the direction perpendicular to it. When accurate dimensional control of the magnets is required, the pole faces of the sintered bodies have to be ground afterwards. More details regarding the manufacturing of ferrite magnets can be found in the reviews of Stäblein [218] and Kools [214].

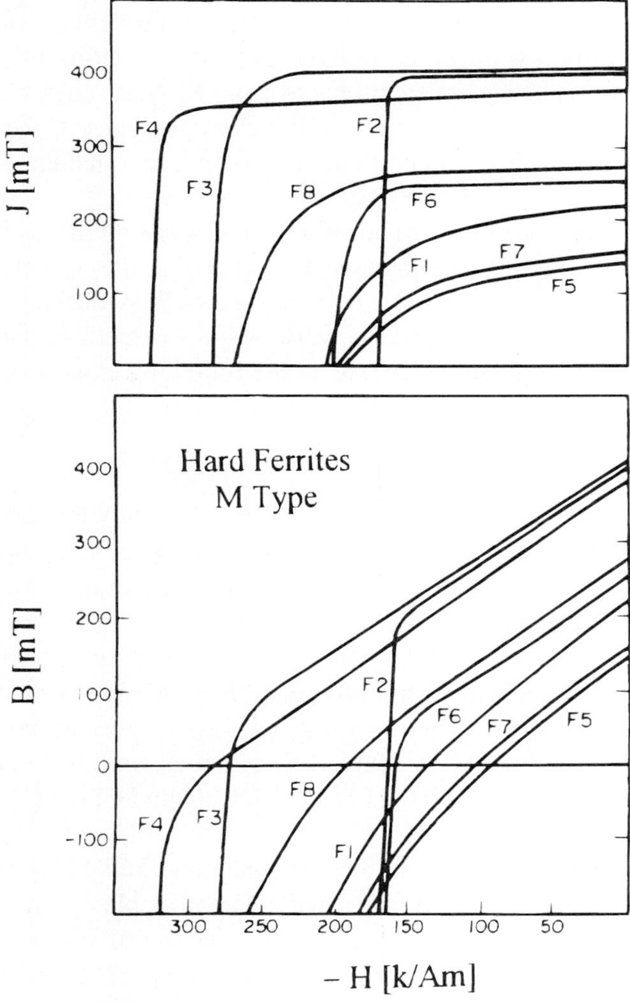

Fig. 20. *Examples of the demagnetizing behaviour of the magnetic polarization J and the corresponding flux density B in permanent magnets prepared from M-type ferrites. From Kools [214].*

Magnets of W type ferrites with the composition $SrZn(LiFe)_{0.5}Fe_{16}O_{27}$ can be prepared in roughly the same way [219]. Calcining of the raw materials is performed at 1330°C, which is slightly higher than in the case of the M-type ferrites. Before annealing (sintering) several types of additives are mixed with the powder obtained from the calcination process. The best results were obtained with SrF_2, although BaF_2 and CaF_2 proved also to be useful. The beneficial influence of the SrF_2 additive was attributed by Kagotani et al. [219] to enhanced formation of W phases due to reaction of the SrF_2 with the α- Fe_2O_3. Compared to M-type ferrites the W-type ferrite magnets are reported to have higher remanence but their coercivity is lower. The M-type ferrites are the more common ones used for the production of permanent magnets. Depending on the degree of sophistication used in the manufacturing process of the magnets there is a whole gamut of different brands and correspondingly different costs. Examples of permanent magnets prepared from M-type ferrites are displayed in Fig. 20. A more detailed description of the characteristics of these magnets and their manufacturing routes can be found in the review of Kools [214].

Generally it can be said that a major purpose of some additives is to control the microstructure of the ultimate magnets by enhancing the particle alignment and reducing the grain growth during prefiring as well as during sintering. Kools [214] has described the influence of various microstructural parameters on the coercivity of ferrite magnets by means of the following expression:

$$_jH_c = \alpha H_A - N_{eff} (J_s + B_r)/ \mu_o \qquad (28)$$

The microstructural parameter α is mainly determined by the grain size whereas N_{eff} depends on the grain shape. Also the fraction of the ferrite phase present in the magnet and the degree of alignment are important since they largely determine the value of the remanence B_r. This expression is of a similar type as discussed already in section 2.3 and is based on nonuniform magnetisation reversal, involving nucleation and propagation of domain walls. A descriptions of the coercivity in terms of uniform magnetization reversal has been proposed by Heinecke [220]. For more detailed discussions of this topic the reader is referred to the reviews of Kools [214] and McCaig and Clegg [191].

The ceramic magnets based on hard ferrite magnets can be classified as low cost - low performance magnets having a wide-spread application. The low cost of the ferrite magnets finds its origin in the very inexpensive raw material and the low manufacturing costs associated with large scale production. This is the main reason that the relative importance of the ferrites is still a considerable one, as may be derived from Fig. 21. The applications comprise anisotropic segments for electric motors, anisotropic rings for loudspeakers and large anisotropic blocks for ore separators. The applications in which the use temperature can become substantially higher than room temperature in particular may profit from the fact that the ceramic magnets have a high chemical stability. Further advantages are that they are electrical insulators, and that the coercivity increases rather than decreases with temperature. The latter property is again beneficial for high-temperature applications even though the temperature coefficient of the coercivity and remanence are undesirably high. It can be derived from the demagnetisation curve shown in Fig. 2 that the low $(BH)_{max}$ value of ceramic magnets is their main disadvantage, restricting their application to magnetic devices in which weight and space are not at a premium.

7. APPLICATION OF PERMANENT MAGNETS

Permanent magnets have become indispensable components in modern technology. They play an important role in many electromechanical and electronic devices used in domestic and professional appliances. For example, an average home contains more than fifty of such devices, and least ten are in a standard family car. Magnetic resonance imaging used as a medical diagnostic tool is an example where large amounts of permanent magnets are used in professional appliances. Permanent magnet materials are furthermore essential in devices for storing energy in a static magnetic field. Major applications involve the conversion of mechanical to electrical energy and vice versa, or the exertion of a force on soft ferromagnetic objects. The applications of magnetic materials in information technology are continuously growing. Apart from the many domestic and professional appliances also automotive and aerospace systems are significant users of permanent magnets, in particular actuators and motion systems.

The most common types of magnets applied at present are alnico type magnets, hard ferrite magnets and rare-earth based magnets (SmCo, NdFeB). Of these the alnico magnets have only a modest coercivity which leads to non-linear demagnetization characteristics, as was discussed in section 4.1. For this reason their applicability is very limited compared to the other two types. The relative commercial importance of these three main categories is illustrated in Fig. 21.

The hard ferrites have higher coercivities than the alnico magnets. Their demagnetizing characteristics are linear but their remanence is fairly low. Consequently the maximum energy product is low, typical values falling in the range 15 - 35 kJ/ m^3. Owing to their low price these magnets are widely applied, although most of the corresponding magnetic devices are rather bulky and often far from optimum performance. Ferrite permanent magnets currently dominate the automotive applications and many of the other applications due to low cost and proven long term stability.

Type of Magnet	T_c (°C)	$(BH)_{max}$ (kJ m^{-3})	B_r (T)	dB_r/dT (%/deg)	$_JH_c$ (kAm^{-1})
Sr-ferrite	450	7-37	0.20-0.44	-0.2	14-300 30-
Alnico	700-860	10-88	0.6-1.4	-0.02	275
Fe-Cr-Co		10-66	1.0-1.6	-0.03	26-51
SmCo$_5$	720	120-200	0.8-1.05	-0.04	600-2000
Sm(Co,Fe,Cu,Zr)$_7$	800	150-240	0.95-1.15	-0.03	450-1300
Nd-Fe-B	310	200-350	1.0-1.3	-0.13	750-1500

Table 4. *Survey of several types of magnets and their room temperature characteristics and Curie temperatures.*

The rare earth based magnets have high values of the coercivity which gives them linear demagnetization characteristics. They have high remanences and typical values of the energy products are 150 kJ/ m^3 for SmCo$_5$, 210 kJ/ m^3 for Sm(Co,Fe,Cu,Zr)$_7$, and 300 kJ/ m^3 for Nd$_2$Fe$_{14}$B. The first two magnet types are expensive owing to the high price of both Sm and Co.

The situation is better for NdFeB magnets. The starting materials are relatively cheap since Nd is cheaper than Sm and Fe is much cheaper than Co, whereas the powder metallurgical processing arts are comparable with those of SmCo$_5$. Here the performance/price ratio for NdFeB is better than for SmCo$_5$. For this reason the market for sintered NdFeB magnets has flourished and is still growing rapidly. The main application areas are small electric motors and actuators where space and low inertia are the main restraints. Prominent examples are voice coil motors in hard disc drives and various types of stepping motors that have become of increasing importance in many applications.

Stepping motors have fairly small sizes and are characterized by their ability of precise incremental position or speed adjustment with electric pulses from digital controllers. They are applied in watches, clocks, timer switches, cameras, etc. The largest applications is in computer peripheral, where they serve, for instance, as floppy-disc head positioners or paper and ribbon advancers. In modern stepping motors the stators consist of the coils system and the rotors are made of rare earth permanent magnets. The advantage of using rare earth permanent magnets is that the high coercivity of these materials makes it possible to axially magnetize the rotor through the thickness with a large number of poles on the circumference. This allows fairly low step angles. More details on electric motors based on sintered NdFeB magnets can be found in surveys written by Howe and Birch [221] and Chalmers et al. [222].

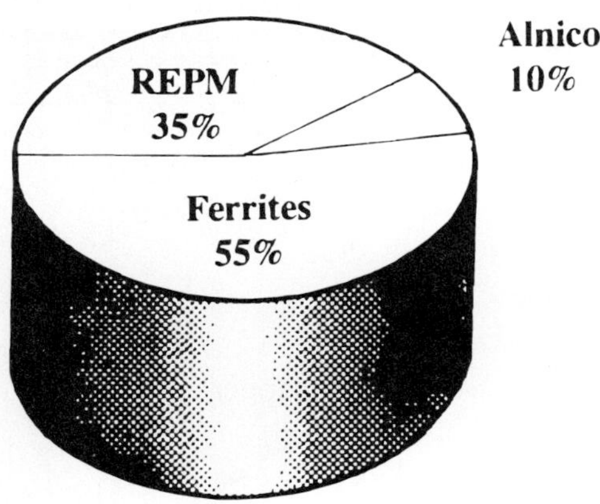

Fig. 21. *Breakdown of the total sales of the three main categories of permanent magnet materials.*

However, sintered NdFeB magnets still have a number of disadvantages. The most important and most fundamental disadvantage is the high temperature coefficient of the coercivity. This property limits the temperature range of operation to about 150°C. For this reason aerospace applications still tend to use sintered Sm-Co based magnets for their elevated temperature capability, high energy product and environmental stability.

The price of sintered NdFeB magnets is far from low and this is due in part to the expensive powder metallurgical route, to the cost of machining and to corrosion protection. The sensitivity to corrosion is not only an extra cost factor by necessitating a protective layer, but it remains an intrinsic weak factor which excludes certain high reliability applications. Also the high values of the remanences and coercivities of the sintered NdFeB magnets are often not required.

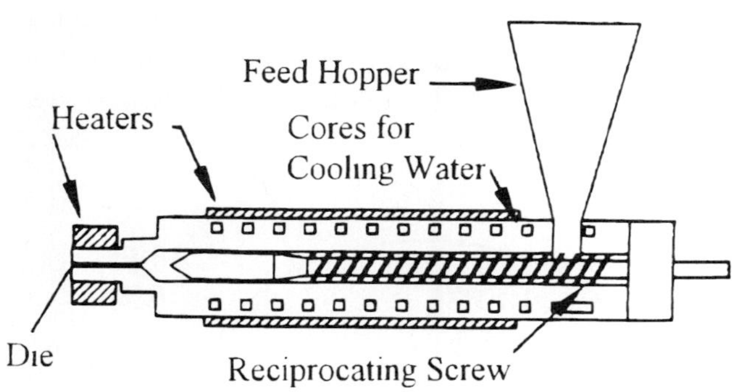

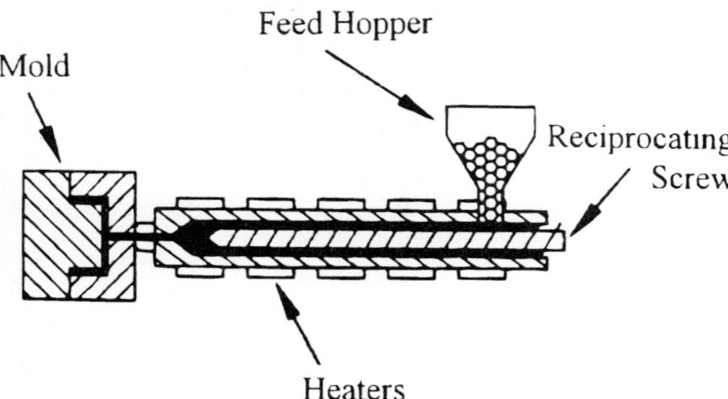

Fig. 22. *Schematic representation of two types of bonded magnet manufacturing, showing (top) extrusion and (bottom) injection molding. From Ormerod and Constantinides [224].*

The increased use of actuators and small electromotors of low weight and low energy consumption (for battery-driven motors in particular) and the increased use of electronically controlled, electrically actuated systems in automobiles (especially for critical vehicle control functions), as well as the increased use of electromagnetic devices in the electronic and electric industry has lead to an enhanced pressure to find improved or alternative permanent magnet materials lying between the low-cost low-performance ferrite magnets and the high-cost high-performance sintered NdFeB magnets.

Bonded NdFeB magnets meet most of the criteria mentioned and their use is growing steadily. For this reason bonded types of permanent magnets will be discussed below in somewhat more detail.

Bonded magnets are generally prepared by blending coercive powder with a binder or by encapsulating coercive powder in a binder, followed by compacting or molding the material to the final shape. Depending on the binder, the magnets produced may be flexible or rigid. Nitrile rubber or vinyl are commonly used for the former type whereas for the latter type nylon, teflon, polyester or thermoset epoxies are employed. There are four main manufacturing routes for bonded magnets: calendering, injection molding, extrusion and compression bonding.

Calendering involves the processing of the material between rollers and leads to continuous strips of adjustable thickness and length. Magnetic loading can be up to 70% by volume.

Injection molding involves forcing the heated mixture of magnetic powder and thermoplastic binder via tubes into a mold where it is allowed to cool and harden. Also in this case magnetic loading can become up to 70%. A schematic representation of this type of manufacturing is shown in Fig. 22.

Extrusion makes use of an orifice through which the compounded material is squeezed, as illustrated in Fig. 22. During the extrusion process the orifice is heated and the profile is controlled as the compounded material cools. The magnetic loading can reach slightly higher values (up to 75% by volume) than in the two previous cases.

Compression bonding uses magnetic powder blended with an epoxy that has been liquefied by dissolution in an organic solvent. The latter is evaporated during the blending process, leaving the magnetic particles in an epoxy-encapsulated state. After drying the coated powder is flowed into a conventional powder press and compacted into the desired shape. Finally, the compacted magnet is cured in an oven. Magnetic loadings up to 80 % can be reached by this process.

Because of the presence of the binder the magnetic remanence is always lower than that of the fully dense magnetic material, as obtained for instance by sintering. There are, however, obvious advantages. The major advantage is that bonded magnets can be produced in net shape. Even fairly complicated shapes may be involved, as in injection molding. Another advantage is their superior mechanical properties. Bonded magnets are mainly manufactured from ferrite powder or coercive NdFeB based powders. Bonded magnets based on NdFeB, owing to their superior intrinsic magnetic properties, offer substantial advantages in terms of size weight and performance over bonded ferrite magnets and even over sintered ferrite magnets.

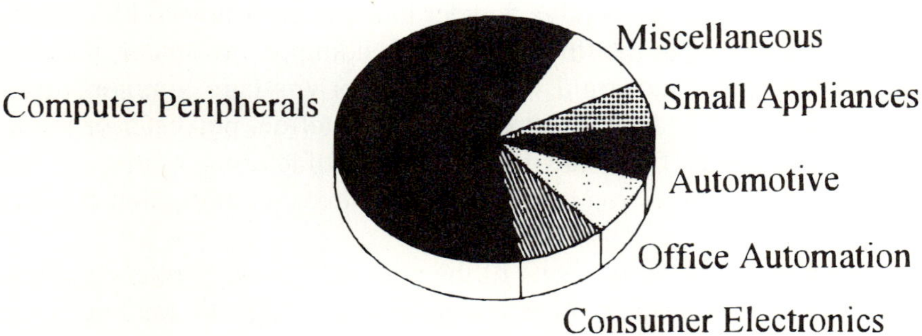

Fig. 23. *Breakdown of applications of bonded magnets based on NdFeB. From Croat et al.[223]*

A breakdown of the application of bonded NdFeB magnets is displayed in Fig. 23. Most of the bonded NdFeB magnets go into computer peripherals where they are mainly used in hard disk drive spindle motors, CD-ROM pick-up and CD-ROM spindle motors. The other applications shown in Fig. 23 include stepper motors used in floppy disc drives, VCRs, camcorders and printers, magnets used in watches, clocks speakers and headphones. For a more detailed discussion of bonded magnet manufacturing and their applications the reader is referred to the review articles of Croat [223] and Ormerod and Constantinides [224].

8. REFERENCES

[1]   W.M. Hubbart, E. Adams and J.V. Gilfrich, J. Appl. Phys. 31 (1960) 368.
[2]   G. Hoffer and K.J. Strnat, IEEE Trans. Mag. 2 (1966) 487.
[3]   K.J. Strnat, G. Hoffer, C.J. Olson, W. Ostertag and J.J. Becker, J. Appl. Phys. 38 (1967)1001.
[4]   W.A.J.J. Velge and K.H.J. Buschow, J.Appl. Phys. 39 (1968) 1717.
[5]   K.H.J Buschow and W.A.J.J. Velge, Z. Angew. Phys. 26 (1969) 157.
[6]   R.K. Mishra,G. Thomas, T. Yoneyama, A. Fukono and T. Ojima, J.Appl. Phys. 52 (19981) 2517.
[7]   M. Sagawa, S. Fujimura, M. Togawa M. and Y. Matsuura Y., J. Appl. Phys. 55 (1984), 2083.
[8]   J Croat., J.F. Herbst, R.W Lee and F.E. Pinkerton, J. Appl. Pys. 55 (1984) 2078.
[9]   D.B. de Mooij and K.H.J. Buschow J. Less-Common Met. 142 (1988) 349.
[10] J.M.D. Coey and H. Sun, J.Magn. Magn. Mater 87 (1990) L251.
[11] K.J. Strnat in Ferromagnetic Materials Vol.4 pp.131, North Holland, Amsterdam (1988), E.P. Wohlfarth and K.H.J. Buschow Eds.
[12] K.H.J. Buschow in Ferromagnetic Materials Vol.4 p.l., North Holland, Amsterdam (1988), E.P. Wohlfarth and K.H.J. Buschow Eds.
[13] H.S. Li and J.D.M. Coey in Handbook of Magnetic Materials, Vol.6, pp3, North Holland, Amsterdam (1991), K.H.J. Buschow,Ed.
[14] K.H.J. Buschow, Repts. Prog. Phys. 54 (1991) 1123.
[15] J.F. Herbst, Rev. Mod. Phys. 63 (1991) 819.
[16] G.J. Long and F. Grandjean Eds., Supermagnets, Hard Magnetic Materials pp.703, Kluwer Acad. Publ., Dordrecht (1991).
[17] K.H.J. Buschow in Materials Science and Technology, Vol. 3B, pp   451, VCH, Weinheim (1994), R.W. Cahn et al. Eds.
[18] H. Fujii and Hong Sun in Handbook of Magnetic Materials, Vol.9, pp 303, North Holland, Amsterdam (1995), K.H.J. Buschow,Ed.
[19] J.D.M. Coey Ed.: Rare Eath Iron Permanent Magnets, Oxford Science Publ., Oxford 1996.
[20] K.H.J. Buschow in Handbook of Magnetic Materials, Vol.10, Elsevier Science, Amsterdam (1997), K.H.J. Buschow Ed.
[21] L. Nordström, M.M.S. Brooks and B Johansson, UOIP 1285, 1991.
[22] G.H.O. Daalderop, P.J. Kelly and M.F.H. Schuurmans, Phys. Rev. B 44 (1991) 12054.
[23] G.H.O. Daalderop, P.J. Kelly and M.F.H. Schuurmans, Phys. Rev. B 53 (1996) 14415.
[24] M.T. Hutchings, Solid State Phys. 16 (1964) 227.
[25] J.J.M Franse and R.J Radwanski in Magnetic Materials, Vol. 7, North Holland, Amsterdam (1993), K.H.J. Buschow Ed.

[26] J.M. Cadogan, J.M.D. Coey, J. Gavigan, D. Givord and H.S. Li J. Phys. F18 (1988) 779.

[27] M. Yamada, H. Kato, H. Yamamoto and Y. Nakagawa, Phys. Rev. 38 (1988) 620.

[28] C. Rudowicz, J. Phys. C18 (1985) 1415.

[29] P.A. Lindgard and O. Danielsen, Phys. Rev. B11 (1975) 351.

[30] W. Sucksmith and J.E. Thomson, Proc. Roy. Soc. London A225 (1954) 362.

[31] P.C.M. Gubbens, A.M. van der Kraan and K.H.J Buschow, Phys. Rev, 39 (1989) 12548.

[32] J.M. Friedt, A. Vasques, J.P. Sanchez, P.L'H,ritier and R. Fruchart, J. Phys. F 16 (1986) 651.

[33] H.H.A. Smit, R.C. Thiel and K.H.J. Buschow, J. Phys. F 18 (1988) 295.

[34] C. Mayer, J.P. Gavigan, G. Czjzek and H.-J. Bornemann, Solid State Comm. 69 (1989) 83.

[35] P.C.M. Gubbens, A.M. van der Kraan, T.H. Jacobs and K.H.J. Buschow, J. Magn. Magn. Mater. 80 (1989) 265.

[36] P.C.M. Gubbens, A.M. van der Kraan and K.H.J. Buschow, Hyperfine Inter. 53 (1990) 37.

[37] F.M. Mulder, R.C. Thiel and K.H.J. Buschow, J. Alloys Compd. 202 (1993) 29.

[38] F.M. Mulder, R.C. Thiel and K.H.J. Buschow J. Alloys Compd. 203 (1994) 97.

[39] R. Coehoorn, K. H. J. Buschow, M. W. Dirken and R. C. Thiel, Phys. Rev. B 42 (1990) 4645.

[40] K.H.J. Buschow, F.M. Mulder and R.C. Thiel and J. Alloys Compds. 275/277 (1998).

[41] K.H.J.Buschow, R. Coehoorn, F.M. Mulder and R.C. Thiel, J. Magn. Magn. Mater., 118 (1993) 347.

[42] F.E. Kayzel, Thesis, University Amsterdam, Amsterdam 1997.

[43] R. Coehoorn in Electron Theory in Alloy Design, The Intitute of Materials (1992), D.G. Pettifor and A.H. Cottrell, Eds.

[44] K. Hummler and M. Fähnle, Phys. Rev. B 53 (1996) 3272.

[45] F.M. Mulder, R.C. Thiel, R. Coehoorn, T.H. Jacobs and K.H.J. Buschow, J. Magn. Magn. Mater. 117 (1992) 413.

[46] R. Coehoorn and K.H.J. Buschow, J. Appl. Phys. 69 (1991) 5590.

[47] M.W. Dirken, R.C. Thiel, R. Coehoorn, T.H. Jacobs and K.H.J. Buschow, J. Magn. Magn. Mater.94 (1991) L 15.

[48] P. Uebele, K. Hummler and M. Fähnle, Phys. Rev. B 53 (1996) 3296.

[49] R. Coehoorn and G.H.O. DaalderopJ. Magn. Magn. Mater. 104-107 (1992) 1081.

[50] M. Bog,, C. Czjzek, D. Givord, C. Jeandry, H.S. Li and J.L. Oddou, J. Phys. F 16 (1986) L 67-

[51]   P.C.M. Gubbens, A.M. van der Kraan and K.H.J. Buschow, Hyperfine
       Inter. 50 (1989) 685.

[52]   R. Fruchart, P. L'H,ritier, P. Dalmas de R,otier, D. Fruchart, P. Wolfers,
       J.M.D. Coey, L.P. Ferreira, R. Guillen, P. Vulliet and A. Yaouanc, J. Phys
       F 17 (1987) 483.

[53]   A. Vasques, J.P. Sanchez, J. Less-Common Met. 126 (1987) 70.

[54]   K.H.J. Buschow, G. Czjzek, H.-J. Bornemann and R. Kmiec, Solid   State
       Comm. 71 (1989) 759.

[55]   M.W. Dirken, R.C. Thiel., T.H. Jacobs and K.H.J. Buschow, J. Less-
       Common Met. 168 (1991) 269.

[56]   X.F. Zhong and W.Y. Ching, Phys. Rev. B39 (1989) 12018.

[57]   L. Steinbeck, M. Richter, U. Nitzsche and H. Eschrig, Phys. Rev. B 53
       (1996) 7111.

[58]   J.P. Liu, F.R. de Boer, P.F. De Ch$f$tel, R. Coehoorn and K.H.J.
       Buschow, J. Magn. Magn. Mater. 132 (1994) 159.

[59]   R. Verhoef, P.H. Quang, J.J.M. Franse and R.J. Radwanski, J. Magn.
       Magn. Mater. 75 (1988) 319.

[60]   M.S.S. Brooks and B. Johansson in "Magnetic Materials Vol. 7" Elsevier
       Science Publ. Amsterdam 1993, K.H.J. Buschow Ed.

[61]   M. Liebs, K. Hummler and M. Fähnle, J. Magn. Magn. Mater. 124,
       (1993) L239.

[62]   M. Liebs, K. Hummler and M. Fähnle, J. Magn. Magn. Mater. 128,
       (1993) 190.

[63]   K. Hummler and M. Fähnle, Phys. Rev. 53 (1996) 3290.

[64]   E.C. Stoner and E.P. Wohlfarth, Philos. Trans. R. Soy. London 240
       (1948) 599.

[65]   H. Kronmüller, K.D. Durst and M. Sagawa, J. Magn. Magn. Mater. 74,
       (1988) 291.

[66]   H. Kronmüller in Proc. NATO-ASI "Supermagnets, Hard Magnetic
       Materials" Kluwer Acad. Publ. Dordrecht (1991), G.J. Long and F.
       Grandjean Eds.

[67]   H. Kronmüller in Proc. 8th Int. Symposium on Magnetic Anisotropy and
       Coercivity in Rare Earth Transition Metal Alloys, Birmingham 1994.

[68]   D. Givord, P. Tenaud and T. Viadieu, IEEE Trans. Magn. MAG-24,
       (1988) 1921.

[69]   D. Givord, O. Lu, M.F. Rossignol, P. Tenaud and T. Viadieu,J. Magn.
       Magn. Mater. 83 (1990) 183.

[70]   T. Schrefl and J. Fidler, J. Magn. Magn. Mater. 111 (1992) 105

[71]   Y. Uesaka, Y. Nakatani and N. Hayashi, J. Magn. Magn. Mater. 123
       (1992) 209.

[72]   T. Schrefl, R. Fischer, J. Fidler and H. Kronmüller J. Appl. Phys. 76
       (1994) 7053.

[73]   T. Schrefl, J. Fidler and H. Kronmüller, Phys. Rev. B, 49 (1994) 6100.

[74] T. Schrefl, J. Fidler and H. Kronmüller, J. Magn. Magn. Mater. 138 (1994) 15.

[75] E.H. Feutrill, L. Folks, P.G. McCormick, P.A.I. Smith and R. Street, 8th Int. Symposium on Magnetic Anisotropy and Coercivity in RE-TM Alloys, Birmingham University 1994, C.A.F. Manwaring et al., Eds, pp. 297.

[76] T. Schrefl and J. Fidler J. Magn. Magn. Mater. 157/158 (1996) 331.

[77] T. Schrefl and J. Fidler, J. Appl. Phys. 79 (1996) 6458.

[78] H. Kronmüller, M. Becher, M. Seeger and A. Zern, Proc. 9th International Symposium on Magnetic Anisotropy and Coercivity in Rare Earth Transitiom Metal Alloys, Vol. 2, pp. 1. World Scientific, Singapore (1996), P.F. Missell et al., Eds.

[79] G. Schneider, E.Th. Henig, G. Petzow and H.H. Stadelmaier Z. Metallkde 77 (1986) 755.

[80] G. Schneider, F.J.G. Landgraf and F.P. Missell, J. Less-Common Met. 153 (1989) 169.

[81] F.J.G. Landgraf, G. Schneider, V. Villas-Boas and F.P. Missell, J. Less-Common Met. 163 (1990) 209.

[82] J.M. Moreau, L. Paccard, J.P. Nozières, F.P. Missell, G. Schneider and V. Villas-Boas, J. Less-Common Met. 163 (1990) 245.

[83] C. Herget in Proc. 8th Int. Workshop on R.E. Magnets and their Applications, Canberra 1992, pp. 60.

[84] D. Fruchart, M. Bacmann, P. de Rango, O. Isnard, S. Liesert, S. Miraglia, S. Obbade, J.-L. Soubeyroux, E. Tomey and P. Wolfers, J. Alloys Comp. 253-254 (1997)121.

[85] P.J. McGuiness, E.J. Devlin, I.R. Harris, E. Rozendaal and J. Ormerod J. Mat. Sci., 24 (1989) 2541.

[86] M. Sagawa, P Tenaud, F. Vial and K Hiraga, IEEE Trans. Magn. MAG-26 (1990) 1957.

[87] S Hirosawa, H. Tomizawa, S. Mino and A. Hamamura IEEE Trans. Magn. MAG-26 (1990) 1360.

[88] S. Hirosawa, S. Mino and H. Tomizawa J. Appl. Phys. 69 (1991) 5844

[89] J.P. Nozières, R. Perrier de la Bâthie and J. Gavinet, J. Phys. (Paris) 49 C8 (1988) 667.

[90] J.P Nozières et al. International Patent WO 91/00602.

[91] K. Akioka, K. Kobayashi, T. Yamagami and T. Shimoda, J. Appl. Phys. 69 (1991) 5829.

[92] W.C. Chang, C.R. Paik, H. Nakamura, N. Takahashi, S. Sugimoto, M. Okada and M. Homma, IEEE Trans Mag. MAG-27 (1991) 3584

[93] R.W. Lee, Appl. Phys. Lett. 46 (1985) 790.

[94] R.W. Lee, E.G. Brewer and N.A. Schafel, IEEE Trans. Magn. MAG-21(1985) 1958.

[95] T. Otsuka and E. Otsuki, US Patent No.5,011,552; European Patent No. 0261 579 B1.

[96] F.M. Ahmed D.S. Edgley and I.R Harris. J. Alloys Comp.209 (1994) 363.

[97] M.H. Gandehari, U.S. Patent No. 5.015.304.

[98] K. Sasaki, T. Otsuka and T. Fujiwara, European Patent No. 0249 973 B1.

[99] A. Hamamura, Y. Kaneko, Y. Okajima, K.Takeja and S. Okada, European Patent 0553 527 A1 (February 1992).

[100] Y. Kaneko,European Patent 0 561 650 A2.

[101] C.H. De Groot, K.H.J. Buschow, F.R. de Boer and C.G.C.M. de Kort, J.Appl. Phys 83 (1998) 388.

[102] M.H. Gandehari, Appl. Phys. Lett. 48 (1986) 548.

[103] M. Velicescu, P. Schrey and W. Rodewald, IEEE Trans. Mag. 31 (1995) 3623.

[104] A. Yan, X. Song and X.. Wang, J.Magn. Magn.Mater. 169 (1997) 193.

[105] O.M. Ragg and I.R. Harris, J. Alloys Comp. 256 (1997) 252.

[106] E.G. Brewer, US Patent No. 453434, European Patent No. A-0133758, European Patent No. 0434 113 A2.

[107] M. Saito et al. European Patent No. 0 392 799 B1.

[108] R.K. Mishra, V. Panchauwthan and J. Croat, J. Appl. Phys. 73 (1993) 6070.

[109] M. Shinoda, K. Iwasaki, S. Tamingawa and M. Tokunaga, Proc. 12th Int. Workshop on R.E. Magnets and their Applications; Canberra 1992, pp.13.

[110] R. Perrier de la Bâthie et al., European Patent No.87903341.

[111] K. Akioka, K. Kobayashi, T. Yamagami and T. Shimoda, US Patent 5,213,631.

[112] H. Kwon, P. Bowen and I.R. Harris, Proc. 12th Int. Workshop on R.E. Magnets and their Applications, Canberra 1992, pp. 705.

[113] C.H. Sellars, D.J. Branagan, T.A. Hyde, L.H. Lewis, J.-Y. Wang and V. Panchanathan, Workshop on High Coercivity Materials, Perth 1997

[114] T. Takeshita et al. US Patent No. 4,981,532.

[115] I.R. Harris, Proc. 12th International Workshop on R.E. Magnets and their Applications, Canberra 1992, pp. 347.

[116] T. Takeshita and R. Nakayama, Proc. 12th International Workshop on R.E. Magnets and their Applications, Canberra 1992, pp. 670.

[117] R. Nakayama and T. Takeshita, J. Alloys Comp. 192 (1993) 259; 192 (1993) 231.

[118] R. Nakayama, T. Takeshita and T. Ogawa, US Patent 5,228,930.

[119] V.A. Yartis, O. Gutfleisch, V.V. Panasyuk and I.R. Harris, J. Alloys Comp. 253-254 (1997) 128.

[120] S. Liesert, D. Fruchart, P. de Rango and J.-L. Soubeyroux, J. Alloys Comp. 253-254 (1997)140; S. Liesert, A. Kirchner, W. Grunberger,

A. Handstein, P. De Rango, D. Fruchart, L. Schultz and K.-H. Muller, J. Alloys Comp. 266 (1998) 260.

[121] J. Eisses, D.B. de Mooij, K.H.J. Buschow and G. Martinek, J. Less-Common Met. 171 (1991) 17.

[122] Y.C. Sui, Z.D. Zhang, W. Liu, Q.F. Xiao, X.Y. Zhao, T. Zhao and Y.C. Chuang, J. Alloys Comp.264 (1998) 228.

[123] K.H.J. Buschow, J. Less-Common Met. 37 (1974) 91.

[124] K.H.J. Buschow, Repts. Prog. Phys. 40 (1977) 1179.

[125] K. Kumar J. Appl. Phys. 63 (1988) R13.

[126] M.F. de Campos, F.J.G. Landgraf, R. Machado, D. Rodrigues, S.A. Romero, A.C. Neiva and F.P. Missell, J. Alloys Comp. 267 (1998) 257.

[127] A.E. Ray and S. Liu, Proc. 12th Int. Workshop on R.E. Magnets and their Applications; Canberra 1992, pp.552.

[128] K.H.J. Buschow F.A.J. den Broeder, J.Less-Common Met. 33 (1973) 191.

[129] B.Y. Wong, M. Willard and D.E. Laughlin, J. Magn. Magn. Mater. 169 (1997)178.

[130] C. Maury, L. Rabenberg and C.H. Allibert, Phys. Stat. Sol.(a) 140 (1993) 57.

[131] A. Lefevre, L. Cataldo, M.Th. Cohen-Adat and B.F. Mentzen, J. Alloys Comp. 255 (1997) 161.

[132] L. Cataldo, A. Lefevre, F. Ducret, M.-Th. Cohen-Adat, C. Allibert and N. Valignat, J. Alloys Comp. 241 (1996) 216.

[133] A. Lefevre, M.Th. Cohen-Adat and B.F. Mentzen, J. Alloys Comp. 256 (1997) 207.

[134] K.H.J. Buschow and A.S. van der Goot, Acta Cryst. B 27 (1971) 1085

[135] S. Liu and A.E. Ray, Proc. 10th Intl. Workshop on Rare Earth Magnets and Their Applications, Kyoto (1989) Japan, p. 265.

[136] S. Liu and A.E. Ray, IEEE Trans. Magn., 25 (1989) 3785.

[137] S. Liu, A.E. Ray and H.F Mildrum., J. Appl. Phys. 69 (1990) 5853.

[138] A.S. Kim J. Appl. Phys. 81 (1997) 5609.

[139] B.M. Ma, Y.L. Liang and C.O. Bounds, J. Appl. Phys. 81 (1997) 5612.

[140] J.C. Koo, IEEE Trans. Magn. MAG-20 (1984) 1593.

[141] K. Akioka, M. Sakurai and T. Shimoda IEEE Trans. Magn., MAG-23 (1987) 2714.

[142] J.M. Cadogan, H. Li, A. Margarian, J.B. Dunlop, D.H. Ryan, S.J. Collocott, R.L. Davis, J.Appl. Phys. 76 (1994) 6138.

[143] C.D. Fuerst, F.E. Pinkerton, J.F. Herbst, J.Magn. Magn. Mater. 129 (1994) L115.

[144] G.C. Hadjipanayis, Y.H. Zeng, A.S. Murthy, W. Gong and F.M. Yang, J. Alloys Comp. 222 (1995) 49.

[146] B.Nasunjilegal, F.M. Yang, N. Tang, W.D. Qin, J.L. Wang, J.J. Zhu, H.Q. Guo, B.P. Hu, Y.Z. Wang , H.S. Li, J.Alloys Comp. 222 (1995) 57.

[147] X.C. Kou, R. GröSssinger, M. Katter, J. Wecker, L. Schultz, T.H. Jacobs and K.H.J. Buschow, J. Appl. Phys. 70 (1991) 2272.

[148] M. Brouha, K.H.J. Buschow and R. Miedema, IEEE Trans. MAG-10 (1974)182.

[149] S.R. Mishra, G.J. Long, O.A. Pringle, D.P. Midlleton, Z. Hu, W.B. Yelon, F. Grandjean, and K.H.J. Buschow, J.Appl. Phys. 79 (1996) 3145.

[150] P. Schobinger-Papamantellos, K.H.J. Buschow, F.R, de Boer, C. Ritter O. Isnard and F. Fauth, J.Alloys Comp.267 (1998)59.

[151] P. Schobinger-Papamantellos, K.H.J. Buschow and C. Ritter, J.Magn. Magn. Mater. 186 (1998) 21.

[152] P. Mohn and E.P. Wohlfarth, J. Phys F 17 (1987) 2421.

[153] J.P. Woods, B.M. Patterson, A.S. Fernando, S.S. Jaswal, D. Welipitya and D.J. Sellmyer, Phys. Rev. B 51 (1995) 1064.

[154] J.P. Liu, K. Bakker, F.R. de Boer, T.H. Jacobs, D.B. de Mooij and K.H.J. Buschow J. Less-Common Met. 170 (1991) 109.

[155] Y.Z. Wang and G.C. Hadjipanayis, J. Appl. Phys. 87 (1990) 375.

[156] T.S. Chin, M.F. Wu, S.K. Chen and J.M. Yao, J. Alloys Comp. 222 (1995) 143.

[157] M. Katter, J. Wecker, L. Schultz and R. Grössinger, J. Magn. Magn. Mat. 92 (1990) L14.

[158] S. Sugimoto, H. Nakamura, M. Okada and M. Homma 12th Int. Work shop on RE Magnets and their Applications, Canberra 1992, pp. 218.

[159] S. Sugimoto, H. Nakamura, M. Okada and M. Homma 12th Int. Work shop on RE Magnets and their Applications, Canberra 1992, pp. 372.

[160] W. Rodewald, B. Wal, M. Katter, M. Veliscescu and P. Schrey, J. Appl. Phys. 73 (1993) 5899.

[161] A. Fukuno, C. Ishizaka and T. Yoneyama, 12th Int. Workshop on R.E. Magnets and their Applications, Canberra 1992, pp. 60.

[162] B. Gebel, M. Kubis and K.-H. Muller, J. Magn. Magn. Mater 174 (1997) L1.

[163] C. Kuhrt, K. Schnitzke and L. Schultz, J. Appl. Phys. 73 (1993) 6026

[164] J. Jakubowicz and M. Jurczyk, J. Alloys Comp. 266 (1998) 318.

[165] A. Kukuno, C. Ishizaka and T. Yoneyama, 12th Int. Workshop on RE Magnets and their Applications, Canberra 1992, pp. 60.

[166] H. Pan, C. Chen, C.S. Wang, X.. Han and F. Yang, J. Magn. Magn. Mater. 179 (1997) 331.

[167] Y.Z. Wang, B.P. Hu, G.C. Liu, H.S. Li, X.F. Han, C.P. Yang and J. Hu, J. Alloys Comp. (1998) Perth Workshopon High Coercivity Materials

[168] K. Kobayashi, T. Iriyama, N. Imaoka, T. Suzuki, H. Kato and Y. Nakagawa, 12th Int. Workshop on R.E. Magnets and their Applications, Canberra 1992, pp. 32.

[169] H. Yamamoto, T. Kumanbara, H. Nishio and S. Suzuki, J. Alloys Comp. 222 (1995) 67.

[170] R. Arlot, K. Machida, P. De Rango, D. Fruchart and G. Adachi, J. Alloys Comp. 275/277 (1998) ICFE 3.

[171] K.H.J. Buschow, J. Magn. Magn. Mater. 100 (1992) 79.

[172] G.C. Hadjipanayis Y.H. Zeng, A.S. Murthy, W. Gong and F.M. Yang, J. Alloys Comp. 222 (1995) 49.

[173] N. Inoue and S. Suzuki, J.Alloys Comp. 222 (1995) 82.

[174] W.C. Chang C.S. Chen and Y.D. Yao J.Alloys Comp. 222 (1995) 87.

[175] E.F. Kneller and R. Hawig IEEE Trans. Magn. MAG-27 (1991) 3588.

[176] W.F. Miao, J.Ding, P.G. McCormick and R. Street, J. Magn. Magn. Mater. 177/181 (1998) 976.

[177] H. Fukunaga and H. Inoue, Jpn. J. Appl. Phys. 31 (1992) 1347.

[178] K. Ohashi, T. Yokoyama, R. Osugi, and Y. Tawara, IEEE Trans. MAG-23 (1987) 3101.

[179] W. Rave, K. Ramst"ck and A. Hubert, J. Magn. Magn. Mater. 183 (1998) 329.

[180] A. Zern, M. Seeger, J. Bauer and H. Kronmüller, J. Magn. Magn. Mater. 184 (1998) 89.

[181] D. Goll, M. Seeger and H. Kronmüller, J. Magn. Magn. Mater. 185 (1998) 49.

[182] N.M.K. Willcox, J.M. Williams, M. Leonowicz, A. Manaf and H.A. Davies, Proc. 13th Intl. Workshop on Rare Earth Magnets and Their Applications, Birmingham, UK (1994), p. 443.

[183] W.F. Miao, J.Ding, P.G. McCormick and R. Street, J. Alloys Comp. 240 (1996) 200.

[184] T. Schrefl and J. Fidler, J. Magn. Magn. Mater. 177/181 (1998) 970

[185] J.J. Croat, J.F. Herbst, R.W. Lee and F.E. Pinkerton, J. Appl. Phys. 55 (1984) 2078.

[186] A Manaf, M. Leonowicz, H.A. Davies and R.A. Buckley J. Appl. Phys. 70 (1991) 6366.

[187] H. A. Davies J. Mag. Mag. Mater. 157/158 (1996) 11.

[188] S. Huo and H.A. Davies, 8th Int. Symposium on Magnetic Anisotropy and Coercivity in RE-TM Alloys, Birmingham University (1994), C.A.F. Manwaring et al. Eds, pp. 155.

[189] H.A. Davies, J.F. Liu and G. Mendoza, Proc. 9th International Symposium on Magnetic Anisotropy and Coercivity in Rare Earth Transitiom Metal Alloys, Vol. 2, pp. 251. World Scientific, Singapore(1996). P.F. Missell et al. Eds.

[190] R.A. McCurrie in "Ferromagnetic Materials" vol.3 p.l07, North Holland, Amsterdam (1982), E.P. Wohlfarth Ed.

[191] M. McCaig and A.G.Clegg, Permanent Magnets in Theory and Practice, Pentech Press, London (1987).

[192] H. Kaneko, M. Homma and K. Nakamura AIP Proc. 5 (1971) 1088.

[193] S. Jin, N.V. Gayle and J.E. Bernardini, IEEE Trans. Magn. MAG-16 (1980) 1050.

[194] S. Jin, G.Y. Chin and B.C. Wonsiewics, IEEE Trans. Magn. MAG-16, (1980) 139.

[195] S. Jin and G.Y. Chin, IEEE Trans. Magn. 23 (1987) 3187.

[196] M. Homma, J. Inst. Metals Seminar (1991), 53.

[197] G.H.O. Daalderop, Kelly P.J. and M.F.H. Schuurmans. Phys. Rev. 44 (1991) 12054.

[198] H. Kaneko, M. Homma and K. Suzuki Trans. JIM 9 (1968) 124.

[199] K. Watanabe, J. Jpn. Inst. Metals 54 (1990) 1284.

[200] Y. Tanaka, N. Kimura, K. Hono, K. Yasuda, T. Sakurai, J. Magn. Magn. Mater. 170 (1997) 289.

[201] K. Watanabe, Materials Trans. JIM, 32 (1991) 292.

[202] P.B. Braun and J.A.Goedkoop, Acta Cryst. 16 (1963) 737.

[203] X.J. Liu, R. Kainuma, H. Ohtani and K. Ishida, J. Alloys Comp. 235 (1996) 256.

[204] N.I. Vlasova, G.S. Kandaurova, Ya.S. Shur and N.N. Bykhanova, Phys. Met. Metallogr. 51 (1981) 1.

[205] J.J. Van den Broek, H. Donkersloot, G. van Tendelo and J. van Landuyt, Acta Met. 27 (1979) 1497.

[206] D.P. Hoydick, E.J. Palmiere and W.A. Soffa, J.Appl. Phys. 81 (1997) 5624.

[207] S. Kojima, T. Ohtani, N. Kato, K. Kojima, Y. Sakomoto, M. Konno, M. Tsukamara and T. Kubo, AIP Conf. Proc. 24 (1974) 768.

[208] T. Ohtani, N. Kato, S. Kojima, K. Kojima, Y. Sakomoto, I. Konno, M. Tsukahara and T. Kubo, IEEE Trans. Magn. MAG-13 (1977) 1328.

[210] L. Pareti, F. Bolzoni, F. Leccabue and A.E. Ermakov, J. Appl. Phys. 59 (1986) 3824.

[211] A. Yamaguchi, J. Japan. Met. Soc. 28 (1989) 422.

[212] V.A.M. Brabers, in "Magnetic Materials" Vol. 8, Ch.3, Elsevier Science, Amsterdam (1997), K.H.J. Buschow Ed.

[213] H. Kojima, in "Ferromagnetic Materials" Vol. 3 Ch. 5, p. 305, North Holland, Amsterdam (1982), E.P. Wohlfarth Ed.

[214] F. Kools, in Encyclopedia of Materials Science and Eng. Vol. 4 p. 2082, Pergamon Press, Oxford (1986), M.B. Bever Ed.

[215] M. Guillot, in Materials Science and Technology Vol. 3B, pp1, VCH, Weinheim(1994), R.W. Cahn et al. Eds.

[216] Y. Teng, B. Lu and J. Yao, J. Appl. Phys. 81 (1997) 5134.

[218] H. Stäblein, in "Ferromagnetic Materials" Vol. 3, Ch. 7, p. 441, North Holland, Amsterdam (1982), E.P. Wohlfarth Ed.

[219] T. Kagotani, N. Abe, M. Okada and M. Homma, Mat. Trans. JIM 31 (1990) 879.

[220] U. Heinecke, Phys. Stat. Sol. 18 (1966) 569.

[221] D. Howe and T.S. Birch in Proc. NATO-ASI "Supermagnets, Hard
      Magnetic Materials" pp.679, Kluwer Acad. Publ., Dordrecht (1991), G.J.
      Long and F. Grandjean Eds.
[222] B.J. Chalmers, A. Sitzia and E. Spooner in Proc. NATO-ASI
      "Supermagnets, Hard Magnetic Materials" pp.703, Kluwer Acad. Publ.,
      Dordrecht (1991), G.J. Long and F. Grandjean Eds.
[223] J.J. Croat, J. Appl. Phys. 81 (1997) 4804.
[224] J. Ormerod and S. Constantinides, J. Appl. Phys. 81 (1997) 4817.